RANGERS IN COMBAT

A LEGACY OF VALOR

From the snowy forests of Upstate New York and the swamps of South Carolina, to the humid streets of Mogadishu and the snowy mountain peaks of Afghanistan, read accounts of some of the most courageous, daring, and vicious ground combat in the annals of U.S. military history.

J. D. Lock

Author of

THE COVETED BLACK AND GOLD:
Training the Modern United States Army Ranger

CHAIN OF DESTINY

Published by Wheatmark™
610 East Delano Street, Suite 104, Tucson, Arizona 85705 U.S.A.
www.wheatmark.com

Publisher's Cataloging-in-Publication
(Provided by Quality Books, Inc.)

Lock, John D.
    Rangers in combat: a legacy of valor / by J.D. Lock.
    p. cm.
    Includes index.
    "From the snowy forests of upper state New York and the humid swamps of South Carolina to the humid streets of Mogadishu and the snowy mountain peaks of Afghanistan, accounts of some of the most courageous, daring and vicious ground combat in the annals of U.S. military history."
    LCCN: 2005927157
    ISBN: 978-1-58736-499-0 (paperback)
    ISBN: 978-1-58736-765-6 (hardcover)

    1. United States. Army--Commando troops--History.
2. United States--History, Military.   I. Title.

UA34.R36L63 2005   356'.167'0973
                QBI05-600058

But where life is more terrible than death,
it is then the truest valour to dare to live.

Sir Thomas Browne,
Religio Medici (pt. XLIV)
(English author; 1605–1682)

To Colonel Ralph Puckett, Jr., a great American and the quintessential Ranger who sets the warrior standard.

And to Jeannie Martin, Mrs. Jeannie Puckett, the epitome of class and the model "Army Ranger wife," in every positive and independent meaning of that phrase.

ACKNOWLEDGMENTS

I AM INDEBTED TO several people in regards to the creation and development of this book:

Sam Henrie, my publisher. It was Sam's idea and support that led to the creation and publication of this work.

Bill Zucker, PhD, editor pro bono, who took my military "lingo" and phraseology and made it more palatable for civilian consumption. Bill championed my writing—all the while pushing me on style and substance.

Raymond Millen, Lieutenant Colonel, U.S. Army, West Point classmate and Army Ranger buddy, editor, who refined my military history and observations.

Photo Acknowledgements: I would like to thank Ms. Kim Lavdano, PAO 75th Ranger Regiment, and Mir Bahmanyar, author of Shadow Warriors, for their support and contributions.

Rangers in Combat is a significantly better work as a result of all their efforts. Thank you.

FOREWORD

NO DOUBT PEOPLE WHO have completed U.S. Army training, as I somehow managed to do three decades ago, will be attracted to *Rangers in Combat*. John Lock's abiding sense of purpose is to tell the story of the Army Rangers with as much detail and knowledge as possible. My hope is this book will be read by a larger audience, especially those whose responsibility it is to decide how to protect the people of the United States and our interests from hostile and dangerous enemies.

Mr. Lock is correct when he says that a study of training and operations—past, present, and future—provides us with an opportunity to be inspired by the "incredible courage, ability, and heroism of the U.S. Army Rangers." Such inspiration is needed when we consider the nature of the challenges our nation faces. And it is worth remembering as we civilians struggle to remain sufficiently grateful for our freedom and safety.

Mr. Lock brings a soldier's sense of honor, loyalty, and determination to his role as a writer. He has studied the record with the intention of illuminating the path ahead. Thoroughly researched, finely written, and historically fascinating, *Rangers in Combat* is filled with riveting stories and thoughtful observations.

Tucked inside these stories are worthwhile lessons even if your day-to-day life does not include military service. These are things learned, remembered, and applied by everyone who has been a Ranger. Every Ranger is taught to acquire the right sense of urgency when given responsibility for a task, pay attention to the details of the work needed to succeed, and apply the techniques of good planning. Most important, every Ranger is given the chance to lead and to follow; nothing teaches you more about being a good leader than following someone who isn't.

Rangers in Combat is an impressive and invaluable contribution, and a powerful tribute to the countless sacrifices of men and women in uniform.

Bob Kerrey

The former U.S. Senator from Nebraska is a Medal of Honor recipient for his actions as a U.S. Navy SEAL commander in the Republic of Vietnam. He is also a U.S. Army Ranger School graduate. Kerrey is currently the president of The New School, a university in New York City.

CONTENTS

INTRODUCTION

"WHAT IS A RANGER?" The term *Ranger* originated as early as thirteenth century England. Eventually, it crossed the Atlantic to the New World to take root in American soil shortly after war started between the Native American Indians and the colonists of the Commonwealth of Virginia. The first documented use—*rainger*—occurred on 22 March 1622.

The Ranger concept, however, goes well beyond just a name, for the U.S. Army Rangers are among the most elite, if not *the* most elite, combat soldiers in the world with a lineage that predates the very birth of this great nation. With rare exception—namely World War I, for no formal or informal Ranger type organization existed then— they have been at the forefront of all American conflicts and wars, "Leading the Way!"

The Rangers's success as warriors is predicated and built upon principles that can be readily comprehended from its singular history. It is that history, that lineage, that tradition constructed by those generations past and present that we honor. It's that dedication, honor, courage, perseverance, blood, and ultimately the personal sacrifice, to which we owe so much.

My goal is nothing less than to demonstrate with this work the incredible courage, ability, and heroism of the U.S. Army Rangers.

"What is a Ranger?" Rangers Lead the Way—RLTW!

Lieutenant Colonel J. D. Lock
U.S. Army (Retired)

AUTHOR'S NOTE

THE EARLY LINEAGE OF the United States Army Ranger is over two hundred years old and "unofficially" began in 1675 with Captain Benjamin Church, considered by many to be the first American Ranger. When one reviews the history of the Rangers, it can be noted that the lineage falls into two distinct periods of time—early and modern. The early years encompass the pre-American Revolution period with Robert Rogers, the American Revolution with Francis Marion, and the American Civil War with John S. Mosby. Though not "formally" trained in accordance with the standards prescribed by today's prestigious U.S. Army Ranger School, the early deeds of Church, Rogers, Marion, and Mosby easily meet the standards and intent of today's modern Ranger. In actuality, they do more than meet the standard. They are the standard as demonstrated by the fact that all four of these Rangers are members of the Ranger Hall of Fame. Consequently, no history of the United States Army Ranger can be complete without them.

For nearly eighty years following the Civil War, the Ranger concept of warfare remained dormant as the Spanish-American and First World wars were waged. Finally, with the advent of the Second World War, the second historical period that can be referred to as that of the "Modern" Ranger began with the formation of Darby's Rangers. This more conventional view of the Rangers, if one can refer to them as being "conventional" at all, encompasses not only World War II, but also the Korean Conflict, the Vietnam Conflict, Desert One (Iran), Grenada, Panama, Somalia, and the Global War on Terrorism.

Some of the vignettes within are edited and expanded extracts from my previously published work, *To Fight With Intrepidity... The Complete History of the U.S. Army Rangers 1622 to Present*, First Edi-

3

tion published by Simon & Schuster, Pocket Books, 1997, Second Edition published by Fenestra Books, 2001. A more detailed account of Ranger history can be found there.

RANGER CREED

U.S. ARMY RANGERS ARE highly trained and motivated professionals who live by a code called The Ranger Creed, created in 1974 by Command Sergeants Major Neil R. Gentry. The creed is sacrosanct; it is a way of life, a guide for how Rangers conduct themselves. It binds the individual—through unswerving loyalty—to his Ranger buddies, his unit, and ultimately to mission accomplishment.

Recognizing that I volunteered as a Ranger, fully knowing the hazards of my chosen profession, I will always endeavor to uphold the prestige, honor, and high *esprit de corps* of the Rangers.

Acknowledging the fact that a Ranger is a more elite soldier who arrives at the cutting edge of battle by land, sea, or air, I accept the fact that as a Ranger my country expects me to move farther, faster, and fight harder than any other soldier.

Never shall I fail my comrades. I will always keep myself mentally alert, physically strong, and morally straight, and I will shoulder more than my share of the task, whatever it may be. One hundred percent and then some.

Gallantly will I show the world that I am a specially selected and well-trained soldier. My courtesy to superior officers, my neatness of dress, and care of equipment shall set the example for others to follow.

Energetically will I meet the enemies of my country. I shall defeat them on the field of battle for I am better trained and will fight with all my might. Surrender is not a Ranger word. I will never leave a fallen comrade to fall into the hands of the enemy, and under no circumstances will I ever embarrass my country.

Readily will I display the intestinal fortitude required to fight on to the Ranger objective and complete the mission, though I be the lone survivor.

Rangers Lead the Way!
(RLTW!)

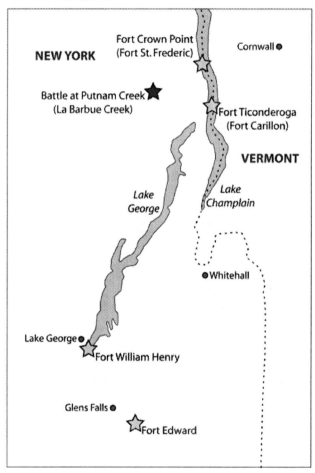

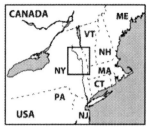

CANADA

ME

VT

NH

NY

MA

CT

PA

USA

NJ

NEW YORK

Fort Crown Point
(Fort St. Frederic)

Cornwall ●

Battle at Putnam Creek ★
(La Barbue Creek)

Fort Ticonderoga
(Fort Carillon)

VERMONT

*Lake
George*

*Lake
Champlain*

● Whitehall

Lake George ●
Fort William Henry

Glens Falls ●

Fort Edward

ONE

LEADERSHIP

Nothing gives one person so much advantage over another as to remain always cool and unruffled under all circumstances.

—Thomas Jefferson
Third U.S. president, 1743–1826

DATE: *15–23 January 1757*

WAR/CONFLICT: *French and Indian War*

LOCATION: *La Barbue Creek (to the French—"Putnam Creek" to the Americans and English), Lake George, New York.*

MISSION: *Reconnaissance and harassment of French, French Canadian and Indian enemy forces in the winter harshness of upper New York State.*

BACKGROUND

THE FRENCH AND INDIAN War lasted from 1754 to 1763. During that period a new, unconventional, and previously untested type of war was executed by the Native American Indian. The British, having seen how successful Ranger style units were against this new and unorthodox style of warfare, began recruiting American frontiersmen to form similar units to serve as auxiliaries of their regular army. The impact of the war upon British infantry techniques and tactics was far-reaching. Impressed by the successful combination of loose-knit Indian fighting and disciplined light-fighting skills, the British Army

9

sought to incorporate this type of unit within their organizational structure.

The Ranger unit that eventually left an indelible mark on American military history and the lineage of the United States Army Ranger, was originally formed as the Ranger Company of the New Hampshire Provincial Regiment under the command of Robert Rogers. As a side note, Rogers was also immortalized in American literature as the main character of Kenneth Roberts' classic novel, *Northwest Passage*.

In February 1755, the governor of New Hampshire commissioned Rogers to the top rank of captain of the 1st Company of a New Hampshire regiment, a group of men that Rogers had personally recruited on his own. Rogers' company was composed of approximately fifty men, all skilled and accomplished in defending themselves and their homes against Indian raids. They were well trained in the back woods, having gained considerable experience from hunting and trapping beaver to hunting and pursuing Indians. Over time, due to these skills, they became known as Rogers' Rangers, the Ranging Company of the Regiment. Rogers, a natural leader, soon set out to make a name for himself and his soldiers.

THE LEGACY

Rogers' winter campaign began on 15 January 1757, with an operation that would eventually transform the company's reputation from that of a scouting corps to that of a fighting organization. Ordered by Major Sparks, the Fort Edward commander, to conduct a reconnaissance of Fort Ticonderoga and Fort Crown Point and "harass the enemy in any way that he saw fit," Rogers departed Fort Edward for Fort William Henry that day with fifty-two Rangers from both his companies.

Arriving at Fort Henry that night, Rogers informed Major Eyre, the fort commander, of the Ranger mission and his requirement to obtain additional Rangers for the patrol. From two additional companies of volunteers commanded by Captains Speakman and Hobbs, Rogers selected Speakman, his older brother, Ensign James Rogers, and thirty additional men.

The next two days were spent preparing for the operation. The uniform and weapons for Rogers' men from Fort Edward were greenish buckskin battledress, individual musket flintlocks or firelocks, scalping knives, and hatchets or tomahawks. Rogers' company officers also carried compasses in the large end of their powder horns, a practice that would soon be emulated by all of the Rogers' Rangers officers. Hobbs and Speakman's Rangers wore gray duffel coats with vests, buckskin breeches, leggings, and water resistant moccasins and socks. They also carried or wore regulation muskets, cartridge boxes, King's regulation shoes, and hatchets—a multi-functional tool that could be used for cutting kindling for fires or for self-defense in close quarters combat.

Ammunition and provisions were issued. Each man was supplied enough powder and balls for sixty rounds. Food rations were carried in a knapsack strapped over a shoulder and consisted of a two weeks's supply of dried beef, sugar, rice, cornmeal, and dried peas. Rum was carried in their wooden canteens. Though Rogers' two Fort Edward companies had come prepared for the winter campaign, the newly acquired volunteers were untrained in the art of winter warfare. Consequently, Rogers' experienced men had to teach them how to make snowshoes for the march. Forming his patrol of eighty-five men on the fort's parade ground on the evening of 17 January, Rogers personally inspected each and every man to ensure that he was properly equipped and provisioned.

Wrapped in blankets—the Ranger winter campaign coat—the patrol moved out in single file, marching the few miles to the first narrows on Lake George. The patrol encamped in an exceptionally defensible position with a steep mountain slope to its rear and the frozen surface of the lake to its front. Sentry posts consisting of six men each were positioned around the perimeter. Two of them were to remain alert at all times. Relief would be done noiselessly and, in the event of needing to sound an alarm, one of them was to retreat silently to notify Rogers in the hope that he could deploy his Rangers quietly to turn the tables on their attackers. Small fires could be built at night but they had to be placed in the heavy part of the woods and in pits three feet deep.

At the break of dawn, the men were roused—the precursor to what would be called "stand-to"—as that was the favorite time for the French and Indians to attack. The still-warm coals from the night-time fires were rekindled, and the Rangers had a hot breakfast of cornmeal gruel washed down with a swig of rum. Rogers made his way around the camp as his men ate, asking about their condition, having noted that several of them had fallen during the evening march. None would admit to any disability, and it was not until the march was started once again that Rogers, standing to one side, was able to identify eleven lame men as the patrol filed past. In spite of their protests, Rogers ordered them to return to Fort William Henry under the charge of one of their number. The stricken Rangers departed, mumbling their protestations as they hobbled away.

Having already ascertained with small reconnaissance parties that the local area was secure, Rogers and his now reduced seventy-four-man Ranger detachment marched within sight of the lake's shoreline in single file, increasing the separation between each man to keep any two from being hit by the same musket ball. An advance guard preceded the main body and flanking parties moved on both sides, at approximately twenty yards. They encamped that second night at a location three miles from Sabbath Day Point.

The following morning, donning their snowshoes, the Rangers moved northwest cross-country through the hills for eight miles to their next camp. By the evening of the 20th, Rogers' force found itself three miles west of Lake Champlain. A few more miles of marching the next morning brought the Rangers to Lake Champlain at Five Mile Point, midway between Fort Ticonderoga and Fort Crown Point, which were five miles away in opposite directions—hence the name.

The weather was bad. Rain and mist reduced visibility. Almost immediately upon their arrival, Rogers' scouts reported two sleighs from Fort Ticonderoga coming toward their position. Ironically, it was from this same location that Rogers had captured two sleighs the previous winter. Rogers quickly organized an ambush. Sending Lieutenant John Stark—who would eventually be considered by many to be as fine a warrior as Rogers—and twenty of Rogers' men farther up, past the point of the shoreline toward Fort Crown Point, Rogers took thirty men with him and moved toward Fort Ticonderoga to cut off the enemy's escape route. Captain Speakman, in the center with his men and positioned where Putnam Creek-La Barbue Creek joined with the lake at a point that jutted out, lay in wait. When the sleighs were opposite his position, Stark was to move out onto the frozen lake to block their movement, thus allowing Speakman to capture them easily.

Unknown to Rogers, however, was a rather significant problem. Earlier that morning, the Fort Ticonderoga commandant had ordered a resupply mission to proceed to Fort Crown Point to load some brandy and hay. A total of ten sleighs, eighty horses, and thirty men set out. The two sleighs and ten men that Rogers' men were preparing to attack were only the advance party of the main force. The two advance sleighs were almost abreast of Speakman's position before Rogers saw the other eight sleighs slowly emerging from the cover afforded by the mist and falling rain.

Rogers quickly dispatched his two fastest men on snowshoes in an attempt to inform his other two sections to execute the ambush when the main body of eight sleighs passed by. Cutting through the woods and successfully gasping out Rogers' orders to Speakman, the two Ranger messengers couldn't reach Stark in time. Unable to see the other eight sleighs because of the bend in the point, Stark and his Rangers sprang from the north side of the lake and, with blood-curdling war cries, spread out on the ice and headed toward the two advance sleighs.

With the ambush sprung, both Speakman and Rogers had no other choice but to join in the attack on the two sleighs. Of the ten accompanying soldiers, seven were captured. The other three, upon seeing and hearing Stark and his men, had jumped on the backs of three horses, cut their traces, and galloped away toward Fort Ticonderoga. Spotting the second group of sleighs, the Rangers removed their snowshoes and gave chase across the ice after the fleeing men and the other eight sleighs. Observing the pack of Rangers bearing down on them, the empty sleighs were able to outdistance Rogers' men quickly and make their way back to the safety of Fort Ticonderoga. Rogers wisely halted the pursuit.

The prisoners were kept isolated from one another and brought before Rogers one at a time. His interrogation revealed that two hundred Canadians and forty-five Indians had just arrived at Fort Ticonderoga with an additional fifty Indians to join them from Fort Crown Point that evening or the following morning. Combined with the three hundred fifty Regulars at Fort Ticonderoga and six hundred Regulars at Fort Crown Point, Rogers realized that there was a formidable force that could soon make him the pursued, rather than the pursuer.

Rogers was caught in a dilemma. He could not just wait while looking for an escape, as it would provide the French an opportunity to gather their forces and catch him between both fortresses. He could not cross the lake and return by way of Wood Creek for fear of being observed from Fort Ticonderoga and subject to ambush. Despite the opposition of some officers, Rogers believed there was only one reasonable recourse left: retreat by the same route in hopes of slipping by Fort Ticonderoga unobserved.

Marching quickly through the snow, the Rangers returned to their previous night's encampment, rekindled the fires, and dried their powder and weapons. Rogers gave orders to Sergeant Walker, commander of the security detachment, to kill their prisoners should the Rangers be attacked in force—a standard practice of that time. Gulping down a hasty meal, the force moved off in single file with its weapons held under blanket-coats to keep them dry in the soaking rain. Rogers led the way with Speakman in the center and Stark bringing up the rear of the main formation. Sergeant Walker followed with the prisoners and a rear guard.

As the Rangers were withdrawing, the angered French were moving to engage. The supply sleigh commander, Major De Rouilly, dispatched a soldier on horse to ride ahead and inform the Fort Ticonderoga commandant, Major De Lusignan, that Rogers had just attacked the supply train from the west. Surmising correctly that Rogers would return west of Fort Ticonderoga through the mountains,

Lusignan dispatched a total of 104 Indians, Regulars, and Canadian volunteers under the dual command of Captains De Basserode and La Granville to intercept them.

Eager to have an opportunity finally to engage Rogers, the French force departed Fort Ticonderoga with only a few rounds of ammunition per man and very little in the way of supplies. A half hour later, De Lusignan dispatched an additional ten men loaded with ammunition and supplies to follow. Basserode's scouts located Rogers' column at around 2 PM, three miles northwest of Fort Ticonderoga. Realizing that the Rangers' route would require them to cross a 75-foot-wide ravine through which Putnam Creek ran, Basserode moved to the location and established a crescent-shaped ambush among the trees and bushes that ran along the crest of the far side gully where it took a turn.

Though a dangerous tactic, Rogers chose to pass through the ravine in an attempt to remain concealed rather than continue along the high ground. This decision resulted in one of the Rangers' bloodiest battles of the war. Having traveled a mile and a half from where they had briefly halted to dry their weapons, the Rangers descended into the ravine. Holding his fire long enough to allow Rogers and the following twelve men to almost reach the summit of the ravine, Basserode's force of 114 men opened fire. The ambush found Rogers caught in the upper kill zone of the ravine—a narrow area inundated with heavy weapons fires and difficult to maneuver through. Speakman with the center section on the floor of the ravine, and Stark's and Walker's rear elements on the opposite hill had yet to enter the ravine.

The vast majority of muskets were probably aimed at Rogers. While half of them misfired because of the rain, the other half amazingly failed to do no more than place one glancing shot across his forehead. Wiping the blood from his eyes, Rogers shouted orders to withdraw back across the ravine. Unfortunately, there were those who could not obey—the initial barrage having killed two and wounded several more.

As the Rangers in the ambush attempted to retire back across the ravine, the attackers, not taking the time to reload after their initial volley, charged down into the ravine with fixed bayonets. Again, it was another miracle that the sections under Rogers and Speakman did not find themselves cut off from the rear. This miracle did not save them from a serious beating, though. Rogers' advance column had just linked up with Speaker's center section when the enemy fell upon them, slashing, stabbing, and shouting. Veteran French Regulars, French Canadians in buckskin, and Indians in war paint joined in the fierce melee within the creek's little ravine.

Rogers' loaded muskets, which had yet to be fired, greeted the cold steel of French and Indian bayonet and tomahawk, but that did not stem the enemy tide. Outnumbered in the ravine nearly two to one, the Rangers heard the welcome volleys of Stark's rear formation from the opposite hill. Having discharged their muskets, Rogers again ordered the advance and center sections to withdraw. The men moved back across the creek and up the gully's side as quickly as they could in their snowshoes. But the going was not easy, for they not only had to fight the enemy that was on their tail, but they were also struggling through four feet of snow. Moving only a few feet at a time up the ravine, a Ranger would have to turn and fight before he could move farther up. The frenzy of the pursuers and the raging survival instincts of the Rangers made for many fierce duels. Captain Speakman fell wounded but managed to conceal himself under a bush as the shouting enemy horde passed him by.

The screams of bayoneted and tomahawked Rangers rose from the ravine floor. With the Rangers below finally spreading out, Stark's men were better able to place supporting fires from above without fear of hitting their own men. Rogers and the survivors made it to Stark's position just as Basserode attempted to envelop the Ranger position with a flanking movement. Upon being informed by Stark of the threat, Rogers dispatched a detachment of marksmen to deal with it. Led by Sergeant Bill Phillips, a noted mixed-blood Indian, the marksmen placed such devastating fire into the flankers that they halted their pursuit abruptly and promptly withdrew back to the main body.

Despite their failure to flank the Ranger position, the French and Indians were still flushed enough with their initial success to attempt a new attack along Rogers' entire front. The enemy slowly worked their way up the ravine, firing from behind trees and bushes until they were within a few yards. Rising en masse, they charged. The high ground and dense cover provided Rogers with the advantage now. The charge failed to advance against what seemed like constant firepower. One of the secrets to Rogers' success was a unique tactic he used to control his fires. Rather than allow all of his men to fire in a single volley, half would discharge their weapons. Then, while they were reloading, the other half would fire, after which the first section would be ready to fire again, having reloaded its rifles. This simple tactic, developed and perfected by Rogers' Rangers, permitted what seemed to be continuous volley fires. The attack dissolved as the French and Indians turned and ran back down the ravine where they established positions from which they could take potshots at the Rangers.

Following a lull in the fighting, the firing soon increased on Rogers' left as another flanking maneuver was attempted. Ensign James Rogers and twelve Rangers were dispatched to counter this threat. Their effective fires soon drove off Basserode's third attack. Having gained some time, Rogers reorganized his defenses, placing Stark in the center, Ensign Brewer on the left, Ensign Rogers and fourteen men on the summit to the rear of the hill, while he took the traditional place of honor on the right side of the formation.

Out of his original force of seventy-four Rangers at the start of the battle, fifty-seven remained. Ten had been killed and seven captured in the ravine. Of the fifty-seven who survived, two were too seriously wounded to fire a weapon. The remaining wounded occupied a place on the line. Security of the seven prisoners was no longer a concern, for they had indeed been executed as per orders and the normal conduct of the day, at the outset of the battle.

Neither Basserode nor his men had any stomach left to try a fourth attack. They had suffered heavy losses in their three previous attempts and could no longer muster an overwhelming force at any point along Rogers' perimeter. Sending off to Fort Ticonderoga for reinforcements, Basserode tried another stratagem and attempted to talk Rogers and his men into surrendering. Calling Rogers by name and flattering him and his Rangers' bravery, Basserode and his officers attempted to convince them that they would be humanely treated if they gave themselves up. If they did not do so, the only alternative was to die when the requested reinforcements arrived. In response, Rogers assured the French that he had plenty of men and supplies left and he himself would "do some scalping and cutting to pieces" if another charge were attempted.

Basserode's request for reinforcements incredibly only brought an additional twenty-six soldiers to his position, bringing the French total to 115 by their account and 250 by Rogers' Rangers' estimate. Bush-fighting warfare continued throughout the afternoon as the French and Indians crawled as close as they dared and traded shots with the Rangers. This sniping was not without its consequences, for Basserode was mortally wounded.

At sundown, Rogers was wounded with a shot along the hand that went through his wrist. Though no longer able to load a musket, Rogers hid the true nature of his wound to ensure the men would not become discouraged. Two other Rangers were seriously wounded with one, Private Joshua Martin, suffering a shattered hip.

Calling a council of war of his officers after sundown, as he was prone to do, Rogers and his officers agreed that the most prudent action to take was to "carry off the wounded of [their] party and take the advantage of the night to return homeward, lest the enemy should send out a fresh party upon them in the morning, [besides, their] ammunition being almost expended [they] were obliged to pursue this resolution."

Gathering in the dark their seven wounded men, the Rangers were relieved to find three arm wounds, two head wounds, and one each in the mouth and side, but none had sustained any wounds to his legs. Under the cover of darkness, Rogers and his men moved off, making fairly good progress. Seeing a fire in the middle of the woods and concerned that it might represent a hostile party, Rogers decided to take a long route around it to place some distance between a potential trouble spot and his exhausted Rangers. The next morning found them on the shore of Lake George, six miles south of the French advance guard positions in the vicinity of the second narrows. Physically drained and unable to advance much farther on foot, the wounded needed to rest.

Despite their fatigued condition, Stark and two others volunteered to march on to Fort William Henry to bring back assistance. Discarding their snowshoes, the three Rangers took to the frozen lake, and despite the effects of the previous long marches, the battle, and their long march during the night, they covered the forty-mile distance to the fort by evening. Exhausted, they stumbled into the fort to inform Major Eyre of the engagement and Rogers' situation.

Sixteen Rangers immediately set forth with sleighs to bring the wounded home. Their journey was not long, for the following morning, 23 January, they linked up with Rogers and his fifty-four survivors as they came staggering through the first narrows of Lake George, wounded in tow.

Soon after Stark's departure, Rogers had reconsidered his position. With the French and Indians so close and Fort William Henry so far, he dared not wait for the sleighs to arrive. Gathering the wounded, the Rangers trudged tiredly after Stark down the lake. Happening to glance behind him, Rogers noticed a dark form following the group. Believing it was a straggler from their column, Rogers sent some men back to get him.

Rather than a straggler from their column, it turned out to be Martin, the private who'd had his hip shattered by a bullet through the stomach. Left for dead on the field of battle, he had recovered himself enough to make off to the woods. There he built a fire to keep from freezing to death. It was his fire that Rogers had skirted and it was this skirting that slowed Rogers down thus allowing Martin, as wounded as he was, to drag himself after his comrades and to overtake them. The moment the Rangers reached him, he collapsed from exhaustion. Much to everyone's surprise, Private Martin survived what appeared to be mortal wounds. Later promoted to sergeant, Martin would ultimately earn his ensign rank in the Ranger Corps.

Despite the Ranger's efforts to ensure that none of their men had been left behind alive after the engagement, others were not so fortunate as Martin. Having concealed himself under a bush during the enemy assault up the ravine, Captain Speakman managed to crawl down the gully and followed the creek out of the battle area. This distance, however, proved to be farther than Rogers' men could look for survivors. Later that day, Speakman was joined by Robert Baker, a British volunteer who had tagged along to see how American Rangers operated, and Private Thomas Brown of Speakman's own company; both men were seriously wounded.

Speakman's and Baker's wounds were so serious they could no longer move. Brown, being the only one still capable of movement, built a fire. Speakman called to Rogers and the Rangers but to no avail. Their only answer was from the enemy. Realizing they could not travel or escape, they decided they would surrender to the French, hoping this could be accomplished prior to being found by the Indians. Just as this decision was reached, Brown observed an Indian moving toward them from over Putnam Creek. Crawling away from the fire, Brown observed the Indian come up to Captain Speakman. Unable to offer any form of resistance because of his wounds, Speakman was stripped and scalped alive. Baker, nearly just as helpless, attempted to pull out a knife and stab himself but he was stopped by the Indian and carried away.

Speakman, still alive, called out to Brown, begging him "for God's sake!" to give him [Speakman] a tomahawk with which he could end his life. Brown could not bring himself to do so and could, in the end, only "[exhort] him as well as I could to pray for mercy, as he could not live many minutes in that deplorable condition, being on the frozen ground, covered with snow." Speakman's final request was to "let his wife know, if [Brown] lived to get home, the dreadful death he died." Following Speakman's demise, Brown attempted to flee the area but was captured. He later managed to escape but was once again captured. His freedom was finally achieved at a later date with an exchange of prisoners.

As it so happened, Speakman was not the only Ranger to be tortured in such a manner. Lieutenant Samuel Kennedy had been mortally wounded and also left on the field of battle. Incapacitated and unable to defend himself, he died under a hail of tomahawks.

Both sides claimed victory. Rogers' Rangers had sustained four-teen killed, nine wounded, and seven captured of seventy-four engaged. The French force of Regulars, Canadians, and Indians num-bered anywhere from 145 to 250 men, depending on whose numbers one is to believe. Either way, Rogers was outnumbered a minimum of two to one. French losses based on their reports—which were al-ways a bit suspect—were put at eighteen killed (to include the seven prisoners) and twenty-seven wounded. When one considers the sig-nificant advantage the French had both in surprise and numbers, it would seem that the Rangers' claim to victory, based upon turning a desperate situation into a well-formulated plan of stubborn defense, had greater validity. While the battle that would be known as the Bat-tle of La Barbue Creek proved to be strategically of no consequence, word of the fight traveled throughout colonial America, raised spir-its, and enriched Rogers' warrior reputation.

OBSERVATION

Military leadership can be defined as "the process of influencing men in such a manner as to accomplish the mission." [FM 22-100, Military Leadership, 29 June 1973, page 1-3, signed by General Creighton W. Abrams, Chief of Staff]. Few leadership traits can be more influential and more conducive to mission accomplishment than those of a com-mander who remains "cool and unruffled" under fire.

The Battle of La Barbue Creek serves as an excellent example of those leadership traits in action. Whether it was maneuvering sec-tions as they fell back through the ambush site, controlling innova-tive volley fires to hold a defensive position against superior enemy forces, or demonstrating exceptional physical perseverance under extremely adverse weather conditions while leading the retreat of a seriously mauled force, Rogers' cool and unruffled leadership style first demonstrated at La Barbue Creek was a proving ground in mak-ing him one of the greatest Ranger leaders in history.

Rogers' personal and aggressive direction throughout the operation also epitomizes another valued leadership trait—the active involvement of the commander. Rogers was in constant motion, checking the status of his men and equipment, shifting his men to threatened areas, ensuring the security of the unit was not relaxed, and exhorting his men to persevere—despite the odds. In this case, Rogers continually adapted to the dynamic situation and maintained the initiative. His successful escape could be ultimately attributed to keeping the enemy off balance.

Robert Rogers

All Rangers are to be subject to the rules and articles of war; to appear at roll call every evening on their own parade ground, each equipped with a firelock, 60 rounds of powder and ball, and a hatchet, at which time an officer from each company is to inspect them to see that they are in order, so as to be ready to march at a minute's warning; and before they are dismissed the necessary guards are to be chosen, and scouts for the next day appointed.

Rule #1
Rogers' Rules of Discipline

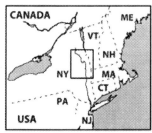

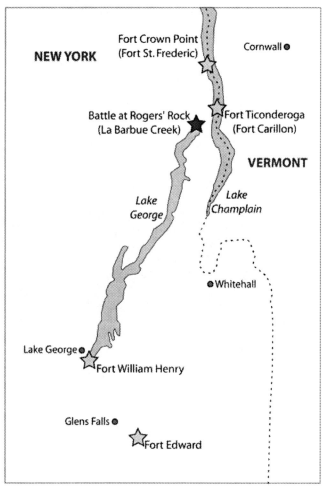

NEW YORK

Fort Crown Point
(Fort St. Frederic)

Cornwall ●

Battle at Rogers' Rock
(La Barbue Creek)

Fort Ticonderoga
(Fort Carillon)

VERMONT

*Lake
George*

*Lake
Champlain*

● Whitehall

Lake George ●

Fort William Henry

Glens Falls ●

Fort Edward

Two

DETERMINATION

Desperate affairs require desperate remedies.

— **Horatio Nelson**
Napoleonic War British Admiral Viscount, 1758–1805

DATE: *10–14 March 1758*

WAR/CONFLICT: *French and Indian War*

LOCATION: *Rogers' Rock (Bald Mountain), vicinity of Fort Ticonderoga, Lake George, New York.*

MISSION: *Combat reconnaissance and ambush of French, French Canadian and Indian enemy forces during a harsh winter.*

BACKGROUND

On 9 January, Captain Robert Rogers had proposed a Trojan horse ruse to fool the enemy. Because Fort Ticonderoga was the most exposed French fort in the region, it remained garrisoned with 350 soldiers during the winter months. Fort Crown Point, however, was farther north and only garrisoned with 150 men. Rogers' proposal was to take four hundred Rangers and march along the backside of the mountains west of Lake George to Lake Champlain. At the lake, they would intercept and capture a group of sleighs traveling along the frozen surface between St. John's and Fort Crown Point. He proposed to outfit some of his French-speaking Rangers with the uniforms of his captives and with the "Trojan Horse" sleighs deceive the

Fort Crown Point garrison commander into opening his gate. Once the gates were open, the remainder of the Rangers concealed nearby would rush the gate and storm the fort before the commandant could learn of the deceit. With Fort Crown Point captured, the umbilical cord of supplies to Fort Ticonderoga would be cut, leaving the fort vulnerable to a siege. Rogers' plan was tentatively approved.

Two months later on 10 March 1758, however, Rogers received from Colonel Haviland, the new garrison commander of Fort Edward, a change to his previous orders. Rather than lead an expedition of four hundred men, Rogers was authorized to lead an element "of 180 men only, officers included."

THE LEGACY

All members of this expedition were volunteers, though Rogers selected only among veterans of his four original companies. As a concession, he allowed three of his five new Ranger companies to be represented by an officer. Each of these three officers, Ensign Andrew Ross, Lieutenant Archibald Campbell, and Ensign Gregory McDonald, were former members of Rogers' Cadet Company and his famous Ranging School.

The expedition included twelve Ranger officers, eleven sergeants, 150 privates of Rogers' Rangers, eight British Regulars, and a corporal of Putnam Connecticut Company. In all, 182 officers and men.

With Rogers at point, the detachment left Fort Edward at mid-afternoon on 10 March and marched only as far as the Half-Way Brook before settling in for the evening. The next day, they reached the first narrows and encamped on the east side of Lake George. Exercising his usual caution, Rogers sent a scouting party three miles farther up the lake on a reconnaissance. Its report was negative. Nevertheless, he posted sentries on the lake and on land during the night.

Up at dawn, they had traveled three miles on the frozen lake when they noticed a dog running across it. Since dogs often traveled with Indians—as they did with Rangers—Rogers established a security perimeter and had a party reconnoiter some islands on the lake. Again the scouts returned with negative reports. Rogers, still cautious, had his force put on snowshoes and keep to the woods until

10 AM when it was opposite Sabbath Day Point. There they remained the rest of the day, keeping a sharp lookout, and only proceeded on the lake later that night. The advance guard was on ice skates ranging out front while Rogers kept the main body grouped tightly to prevent a break in contact. A detachment flanked them to the west, close to shore.

Approximately eight miles from the French advance guard at Coutre Coeur, a campfire was reported on the east shore. Advancing to attack, the Rangers could find no evidence of a fire. The advance scouts were quite possibly deceived by a patch of bleached snow or rotted wood that sometimes turned green/yellow phosphorescence and fooled the best of scouts. Returning to the west bank where they had left their packs, they remained there for the night.

On the morning of the 13th, Rogers held a council of war with his officers. They decided to proceed on with snowshoes and to travel behind Bald Mountain in order to maintain a safe distance away from the French advance guard posts. From 7 to 11 AM, they marched until they were opposite Coutre Coeur. Halting on the back side of the ridge and fully intent on an ambush, they nibbled at a cold meal and waited for the enemy Trout Brook patrol that passed by daily on its return to Fort Ticonderoga.

Deciding not to wait, Rogers and his men resumed their trek down the valley of Trout Brook at 3 PM with a frozen brook close on their left and Bald Mountain—soon to be renamed Rogers' Rock—to their right. With the snow four feet deep even with snowshoes, the walking was tough. A mile and a half later the advance guard report came: "Enemy in view, approximately ninety-five, mostly Indians. "

The enemy was making its way up the valley from the north, along the frozen creek, passing to the west of Rogers, within seventy-five yards of his position. Assuming this to be the main body of the enemy, Rogers moved forward and formed an ambush within a few yards of the bank and waited. With the enemy patrol's leading element opposite their flank, the Rangers sprang the ambush. About forty Indians were killed in the initial volley, with the remainder falling back in disarray. Rogers ordered his Rangers to pursue, sending Captain Bulkeley and his section after the fleeing enemy down the draw. Those remaining behind worked on scalping the fallen Indians. Circumstances were, however, against Rogers and his men. Within fifteen minutes, unforeseen events unfolded that would transform the Rangers' feelings from the ecstasy of victory to the agony of defeat.

Unbeknown to Rogers, on the previous evening, a force of two hundred Indians and thirty-plus French Canadians commanded by Ensign Sieur La Durantaye had arrived at Fort Ticonderoga. The next morning, while Rogers and his men were marching toward Trout Brook, the Indians applied to the fort commandant and received provisions and brandy.

Returning to their camp, they broke open the liquor. One, claiming to be a witch doctor, consulted the "spirits," who revealed to him that there was an English war party about, and not too far away. Not long after six Indians who had been scouting the foot of the lake confirmed the witch doctor's vision: They had seen a large number of fresh tracks where Rogers' detachment had left the lake.

Unfortunately for Rogers and his force, Langy De Montegron was resting at Fort Ticonderoga between raids on Fort Edward. Unable to resist the call to action, Montegron gathered a force of fifty French Canadian and Regular volunteers. Along with the Indians, a total force of 290 Regulars and French Canadians were preparing to move against Rogers. It was La Durantaye's advance guard of ninety-five that Rogers had ambushed. Just minutes behind were another 195 Indian warriors and soldiers.

On hearing firing just ahead, Montegron deployed his force from its on-the-march single-file formation into a bush-fighting extended on-line formation and quickly moved forward. Soon, La Durantaye and his fifty-four survivors rejoined Montegron. Moments later, Bulkeley and his men, still pursuing fleeing Indians, ran right into the opening volley of the French and Indian muskets. Unaware that there was any other force in the area, the Rangers were taken completely by surprise. The French and Indians poured on withering fire in a furious attack. Bulkeley and three of his officers were killed instantly along with nearly fifty of his men. Lieutenant Increase Moore and Ensign McDonald, though both mortally wounded, managed to rally the few survivors who withdrew to Rogers' location.

Rogers' force was scattered. The remnants of his pursuit party were falling back while others were still scalping Indians killed by the Rangers in their ambush. Unable to form a strong defensive front against Montegron's and La Durantaye's attacking forces, Rogers soon found his detachment threatened with envelopment. Ordering his men to fall back to their original position on the slope of Bald Mountain, the Rangers did so but not before another ten men died.

Rogers' men returned fire and fought with a ferocity that forced the enemy to withdraw. The French and Indians rallied again to attack along the front and to the sides. Only the mountain to their back prevented the Rangers from being encircled. Drunk and seeking revenge for their recently scalped comrades—as well as the loss of one of their prominent war chiefs in the ambush—the Indians ferociously pressed their attack.

Rogers had now fewer than 120 men, outnumbered by the enemy two to one. Montegron and La Durantaye capitalized on the Indian intoxication and rage, sacrificing them to the Rangers' fire in hopes of wearing Rogers' men down by attrition. The Rangers held, however, as the Indians attacked, and the Ranger marksmen took their toll.

Repulsed for a third time, the enemy rallied and attacked again forty-five minutes later. With skirmishes to the front, Montegron and La Durantaye attempted to envelop the flanks by sending the Indians on the left and the Canadians on the right. Attacking first, the Indians pressed hard on the Ranger flank. Rogers continued to send small parties of Rangers to reinforce the sector. Fighting was furious and from a distance of only twenty yards. Friend and foe, at times, found themselves intermixed but the Ranger line continued to hold.

Evening was fast approaching and Montegron and La Durantaye knew that if they were to finish Rogers off, it had to be before darkness set in. A final, simultaneous, and unrelenting assault from all three sides was launched. Having lost eight officers and one hundred men, Rogers only had Ensign Joseph Waite and thirty-one Rangers left to hold his center, his newly promoted Indian Ensign Bill Phillips and eighteen men to hold his right, and Lieutenant Edward Crofton,

Captain-Lieutenant Pringle, and Lieutenant Roche—the last two being British officers Haviland had sent—with twenty-two men to secure the Ranger left flank. In all, seventy-six men were left from the
original 182.

The right flank was the first to be enveloped. Surrounded by the
bulk of the Indians, Phillips and his men were quickly losing ground.
Rogers and Waite managed to hold the center until they were down
to twenty-one defenders. Finally, separated from Phillips, nearly
fighting hand to hand and in perilous danger of being totally cut off,
Rogers ordered the center group to break contact with the enemy
and make its way to Crofton and Pringle's position, which was still
relatively intact on the left flank.

As Rogers was withdrawing to the left flank, Phillips on the
right flank was surrendering to La Durantaye. Encircled by an overwhelming number of Indians and promised humane treatment, Phillips and his small band of survivors capitulated. Despite the strong
assurances of humane treatment, all of the prisoners were tied to
trees, slowly tortured, then hacked to pieces when several fresh Indian scalps were found in the coats of slain Rangers. Phillips proved
to be the lone survivor of this group. While the Indians were busy
mutilating and torturing other captives, he was able to free a hand,
secure a knife from his pocket, open it with his teeth, cut his deerskin
cord bond, and escape, only to be captured again. Incredibly, rather
than butcher him on the spot, he was taken to the Indian village of
Sault St. Louis, near Montreal, where once again he escaped, making
his way, ultimately, to the colonies and to freedom.

Linked with the remnants of his left flank and realizing that his
right flank had surrendered, Rogers knew the fight was over and that
the only honorable thing left to do was to save as many of his men as
possible. Implementing one of his Ranging Rules [see end of chap-
ter], Rogers ordered his men to disperse and for each to take a different route to the rendezvous point, the southern end of Rogers' Rock,
where they had concealed their hand sleds on Lake George.

The approaching darkness facilitated their escape; nevertheless,
several of the retreating Rangers were intercepted and captured. Because they were unfamiliar with the terrain, Rogers offered Pringle

and Roche, two surviving British Regulars, a sergeant to guide them through the mountains to the rendezvous point, but they declined the offer. Having damaged their snowshoes during the fight, both realized they would be easy prey for any pursuers and a burden to any Ranger. Making their way from the field of battle under the cover of darkness, the two officers met Rogers' orderly (an aide), who promptly got them lost. On the 19th the orderly died from the cold. The next day, the British officers gave themselves up as they came within sight of Fort Ticonderoga. Ultimately, their very survival was actually dependent on a footrace between the French officers who would take them prisoner and Indians who would butcher them. The Frenchmen won.

On his own now, Rogers' escape was to become the stuff of legend. Discarding his green jacket that also contained his 24 March 1756 commission—which for a while led the Indians and French to believe he was among the dead for surely he would not leave so precious and treasured a document behind—Rogers began to climb the west slope of Bald Mountain. At the summit, he looked down the sheer smooth wall of rock that formed the eastern slope that ran to the frozen surface of Lake George more than one thousand feet below. Behind and below him, he could hear his pursuers shouting excitedly as they located his trail in the moonlight. Conceiving an idea to make those following believe he had gone over the summit to slide down the slope, Rogers loosened the thongs of his snowshoes and turned about-face without moving them. Then, having laced them back up, he proceeded to backtrack along his trail for some distance until he was able to swing himself by a branch off the trail and into a defile, thus leaving no telltale indications that he'd departed the pathway.

The Indians eventually reached the summit just as Rogers was making his way down to the lake, having followed the defile down the mountain. Convinced that he had slid down the slope and that he was being watched over, the Indians did not continue the pursuit. Rogers, the *Wobi Madaondo*, "White Devil," led a charmed existence, in addition to being tactically brilliant.

It was approximately 8 PM when Rogers began to move across the

lake's frozen surface. Within a short time, he encountered other survivors and several wounded men whom they carried to the rendezvous point. From there, Rogers dispatched three Rangers on skates to
move with all haste to Fort Edward and bring back reinforcements,
for Rogers still expected a pursuit. These skaters were soon followed
by four injured Rangers, each strapped in a hand sleigh with two
Rangers pulling each sleigh. Remaining behind with the handful of
survivors, Rogers waited for others to arrive. With no blankets and
unwilling to chance lighting a fire, the small group nearly froze to
death as they huddled through the night. Their lingering was not in
vain, though, for the next morning a few more Rangers wearily staggered in, some wounded. Under Rogers' care, the group started up
the lake to Fort Edward.

The three ice-skating messengers arrived at Fort Edward about
noon on the 14th. Captain Stark and all the Rangers that could be
spared were dispatched to assist the survivors. Encountering Rogers at Sloop Island, about six miles from the head of Lake George,
Stark remained there with Rogers for the night, sending back for
three horse sleighs to carry the wounded. With the sleighs arriving
the next morning, small groups of Rangers began to enter Fort Edward around 3 PM Rogers was the last to arrive at 5 PM, bringing up
the rear.

The numbers were not pretty. Rogers had brought back fifty-two
survivors, eight of whom were seriously wounded. A total of 122 officers and men had been lost on the field of battle. Of those 122, only
one was a prisoner, Ensign Phillips. The remaining 121 were dead.
The French accounts are difficult to accept at face value for La Durantaye claimed to have suffered only eight Indians killed. How does
one reconcile that with the forty men Rogers stated to have killed in
the ambush? All told, it is probably safe to claim that the Rangers
killed at least eighty and wounded a similar number.

Despite the totality of the defeat, General Howe, the new commander of the New York front was extremely impressed Rogers'

detachment, surprised and wounded as it was, still repulsed two assaults by a significantly superior force. Provincial papers applauded their valiant stand, and the incredible battle only served to enhance Rogers' fame even more.

OBSERVATION

When a mission is going well, one does not need a great deal of determination to complete it. However, when a mission is not going well, determination is critical to not only personal survival, but more critically, unit survival as well. Determined to make the best of what he faced—outmanned by enemy forces—Rogers had to prove that he was the superior leader against his nemesis, French Canadian Langy De Montegron. There was no doubt that Rogers' affairs were deeply troubling and required a "desperate remedy" if he and his command were to fight another day. Even losing eight of his twelve Ranger officers, Rogers was still able through sheer determination to save himself and many of his men, producing at a minimum a moral victory from what should have been not only an overwhelmingly catastrophic defeat, but also Rogers' personal demise.

The true test of a commander's skill and determination is when the fate of a battle and his command hangs in the balance. Few are up to the challenge and surrender to their unfortunate fate. As with truly gifted commanders, Rogers refused to acknowledge the inevitable and persevered. As the enemy vise closed in on the remnants of Rogers' command, he defied the conventional military logic of the time and ordered his men to exfiltrate in small numbers to a pre-determined rendezvous point, thus saving his command. With these surviving veterans, Rogers was able to rebuild his Ranger unit and carry on the fight. No matter how many close scrapes Rogers experienced, he was never whipped and always returned like the mythological Phoenix rising from its ashes.

THE 'TRUTH' BEHIND ROGERS' RANGERS STANDING ORDERS

The Ranger Handbook, SH 21-76, published by the Ranger Training Brigade of the United States Army Infantry School, states the following, in part, in regards to the Standing Orders of Rogers' Rangers:

> Ranger techniques and methods were an inherent characteristic of the frontiersmen in the colonies, but Rogers was the first to capitalize on them and incorporate them into a permanently organized fighting force. His "Standing Orders" were written in the year 1759. Even though they are over 200 years old, they apply just as well to Ranger operations conducted on today's battlefield as they did to the operations conducted by Rogers and his men.

Of the three sentences in that quote, the first is true, the second is false, and the third, based on the second, is inherently misleading.

On 14 September 1757, Rogers' Ranging School was officially authorized. Its first group of students was British Cadet volunteers. To structure his training, Rogers drafted twenty-eight tactical rules which came to be known as "Rogers' Rules of Discipline." In 1765, he would have them published as part of his French and Indian War Journals. His rules were detailed, comprehensive, and exceptionally insightful for the period. So insightful were they that they are still largely applicable to the modern battlefield. The "Rules of Discipline" were truly a brilliant discourse on unconventional scouting and skirmishing; they probably constitute the first military field manual written on the North American continent.

So, then, where did the more succinct and entertaining "Roger's Standing Orders" come from? As noted previously, Robert Rogers served as the role model for Kenneth Roberts' protagonist in the 1936 novel *Northwest Passage*. In the conversation between the characters Langdon Towne and Sergeant McNott, in which McNott is explaining to Towne what a Ranger must know, one can find the foundation for the wording of what were to become the "Standing Orders." There can be no doubt that the fictional conversation was predicated on Robert Rogers' Rules of Discipline. However, it *was* a fictional conversation.

This passage from the novel, however, apparently struck a cord with an officer assigned to The Infantry School as a military writer for the 1960 version of Field Manual (FM) 21-50, Ranger Training and Ranger Operations. Within this FM was an appendix on Ranger History that included a paraphrased version of the novel's passage "Standing Orders." A year or two later, a review of the reprinted Journals of Major Robert Rogers by The Infantry School led the staff to question the authenticity of "Rogers' Standing Orders." Despite an attempt on the part of the school to clarify the record about the fictional origin of Rogers' Standing Orders, their efforts were in vain and fruitless. Rogers' Standing Orders became part of lore and legend.

ROGERS' RULES OF DISCIPLINE

1. All Rangers are to be subject to the rules and articles of war; to appear at roll call every evening on their own parade ground, each equipped with a firelock, sixty rounds of powder and ball, and a hatchet, at which time an officer from each company is to inspect them to see that they are in order, so as to be ready to march at a minute's warning; and before they are dismissed the necessary guards are to be chosen, and scouts for the next day appointed.

2. Whenever you are ordered out to the enemy's forts or frontiers for discoveries, if your number is small, march in single file, keeping far enough apart to prevent one shot from killing two men, sending one man or more forward, and the like on each side, at a distance of 20 yards from the main body, if the ground you march on allows it, to give the signal to the officer of the approach of an enemy, and of their number, etc.

3. If you march over marshes or soft ground, change your position

and march abreast of each other to prevent the enemy from tracking you (as they would do if you marched in single file) until you get over such ground, and then resume your former order and march until it is quite dark before you encamp. Camp, if possible, on a piece of ground that gives your sentries the advantage of seeing or hearing the enemy at considerable distance, keeping half of your whole party awake alternately through the night.

4. Some time before you come to the place you would reconnoiter, make a stand and send one or two men in whom you can confide to seek out the best ground for making your observations.

5. If you have the good fortune to take any prisoners, keep them separate until they are examined, and return by a route other than the one you used going out so that you may discover any enemy party in your rear and have an opportunity, if their strength is superior to yours, to alter your course or disperse, as circumstances may require.

6. If you march in a large body of 300 or 400 with a plan to attack the enemy, divide your party into three columns, each headed by an officer. Let these columns march in single file, the columns to the right and left keeping 20 yards or more from the center column, if the terrain allows it. Let proper guards be kept in the front and rear and suitable flanking parties at a distance, as directed before, with orders to halt on all high ground to view the surrounding ground to prevent ambush and to notify of the approach or retreat of the enemy, so that proper dispositions may be made for attacking, defending, etc. And if the enemy approaches in your front on level ground, form a front of your three columns or main body with the advanced guard, keeping out your flanking parties as if you were marching under the command of trusty officers, to prevent the enemy from pressing hard on either of your wings or surrounding you, which is the usual method of savages if their number will allow it, and be careful likewise to support and strengthen your rear guard.

7. If you receive fire from enemy forces, fall or squat down until it is over, then rise and fire at them. If their main body is equal to yours, extend yourselves occasionally; but if they are superior, be careful to support and strengthen your flanking parties to make them equal with the enemy's, so that if possible you may repulse them to their main body. In doing so, push upon them with the greatest resolve, with equal force in each flank and in the center, observing to keep at a due distance from each other, and advance from tree to tree, with one half of the part ten or twelve yards in front of the other. If the enemy pushes upon you, let your front rank fire and fall down, and then let your rear rank advance through them and do the same, by which time those who were in front will be ready to fire again, and repeat the same alternately, as occasion requires. By this means you will keep up such a constant fire that the enemy will not be able to break your order easily or gain your ground.

8. If you force the enemy to retreat, be careful in pursuing them to keep out your flanking parties and prevent them from gaining high ground, in which case they may be able to rally and repulse you in their turn.

9. If you must retreat, let the front of your whole party fire and fall back until the rear has done the same, heading for the best ground you can. By this means you will force the enemy to pursue you, if they pursue you at all, in the face of constant fire.

10. If the enemy is so superior that you are in danger of being surrounded, let the whole body disperse and every one take a different road to the place of rendezvous appointed for that evening. Every morning the rendezvous point must be altered and fixed for the evening in order to bring the whole part, or as many of them as possible, together after any separation that may occur in the day. But if you should actually be surrounded, form yourselves into a square or, in the woods, a circle is best; and if possible make a stand until darkness favors your escape.

11. If your rear is attacked, the main body and flanks must face about the right or left, as required, and form themselves to oppose the enemy as directed earlier. The same method must be observed if attacked in either of your flanks, by which means you will always make a rear guard of one of your flank guards.

12. If you determine to rally after a retreat in order to make a fresh stand against the enemy, by all means try to do it on the highest ground you come upon, which will give you the advantage and enable you to repulse superior numbers.

13. In general, when pushed upon by the enemy, reserve your fire until they approach very near, which will then cause them the greater surprise and consternation and give you an opportunity to rush upon them with your hatchets and cutlasses to greater advantage.

14. When you encamp at night, fix your sentries so they will not be relieved from the main body until morning, profound secrecy and silence being often of the most importance in these cases. Each sentry, therefore, should consist of six men, two of whom must be constantly alert, and when relieved by their fellows, it should be without noise. In case those on duty see or hear anything that alarms them, they are not to speak. One of them is to retreat silently and advise the commanding officer so that proper dispositions can be made. All occasional sentries should be fixed in a like manner.

15. At first light, awake your whole detachment. This is the time when the savages choose to fall upon their enemies, and you should be ready to receive them.

16. If the enemy is discovered by your detachments in the morning, and if their numbers are superior to yours and a victory doubtful, you should not attack them until the evening. Then they will not know your numbers and if you are repulsed your retreat will be aided by the darkness of the night.

17. Before you leave your encampment, send out small parties to scout around it to see if there are any signs of an enemy force that may have been near you during the night.

18. When you stop for rest, choose some spring or rivulet if you can, and dispose your party so as not to be surprised, posting proper guards and sentries at a due distance, and let a small party watch the path you used coming in, in case the enemy is pursuing.

19. If you have to cross rivers on your return, avoid the usual fords as much as possible, in case the enemy has discovered them and is there expecting you.

20. If you have to pass by lakes, keep at some distance from the edge of the water, so that, in case of an ambush or attack from the enemy, your retreat will not be cut off.

21. If the enemy forces pursue your rear, circle around until you come to your own tracks and form an ambush there to receive them and give them the first fire.

22. When you return from a patrol and come near our forts, avoid the usual roads and avenues to it; the enemy may have preceded you and laid an ambush to receive you when you are almost exhausted with fatigue.

23. When you pursue any party that has been near our forts or encampments, do not follow directly in their tracks, lest you be discovered by their rear guards who, at such a time, would be most alert. But endeavor, by a different route, to intercept and meet them in some narrow pass, or lie in ambush to receive them when and where they least expect it.

24. If you are to embark in canoes, or otherwise by water, choose the evening for the time of your embarkation, as you will then have the whole night before you to pass undiscovered by any enemy

parties on hills or other places that command a view of the lake or river.

25. In paddling or rowing, order that the boat or canoe next to the last one wait for it, and that each wait for the one behind it to prevent separation and so that you will be ready to help each other in any emergency.

26. Appoint one man in each boat to look out for fires on the adjacent shores, from the number and size of which you may form some idea of the number that kindled them and whether you can attack them or not.

27. If you find the enemy encamped near the banks of a river or lake that you think they will try to cross for their security when attacked, leave a detachment of your party on the opposite shore to receive them. With the remainder, you can surprise them, having them between you and the water.

28. If you cannot satisfy yourself as to the enemy's number and strength from their fires and the like, conceal your boats at some distance and ascertain their number by a patrol when they embark or march in the morning, marking the course they steer, when you may pursue, ambush, and attack them, or let them pass, as prudence directs you. In general, however, so that you may not be discovered at a great distance by the enemy on the lakes and rivers, it is safest to hide with your boats and party concealed all day, without noise or show, and to pursue your intended route by night. Whether you go by land or water, give out patrol and countersigns in order to recognize one another in the dark, and likewise appoint a station for every man to go to in case of any accident that may separate you.

ROGERS' RANGERS STANDING ORDERS

1. Don't forget nothing.

2. Have your musket clean as a whistle, hatchet scoured, sixty rounds powder and ball, and be ready to march at a minute's warning.

3. When you're on the march, act the way you would if you was sneaking up on a deer. See the enemy first.

4. Tell the truth about what you see and what you do. There is an army depending on us for correct information. You can lie all you please when you tell other folks about the Rangers, but don't ever lie to a Ranger or officer.

5. Don't ever take a chance you don't have to.

6. When we're on the march we march single file, far enough apart so one shot can't go through two men.

7. If we strike swamps, or soft ground, we spread out abreast so it's hard to track us.

8. When we march, we keep moving till dark, so as to give the enemy the least possible chance at us.

9. When we camp, half the party stays awake while the other half sleeps.

10. If we take prisoners, we keep 'em separate till we have had time to examine them, so they can't cook up a story between 'em.

11. Don't ever march home the same way. Take a different route so you won't be ambushed.

12. No matter whether we travel in big parties or little ones, each

party has to keep a scout 20 yards ahead, 20 yards on each flank, and 20 yards in the rear so the main body can't be surprised and wiped out.

13. Every night you'll be told where to meet if surrounded by superior force.

14. Don't sit down and eat without posting sentries.

15. Don't sleep beyond dawn. Dawn's when the French and Indians attack.

16. Don't cross a river by a regular ford.

17. If somebody's trailing you, make a circle, come back onto your own tracks, and ambush the folks that aim to ambush you.

18. Don't stand up when the enemy's coming against you. Kneel down, lie down, hide behind a tree.

19. Let the enemy come till he's almost close enough to touch. Then let him have it and jump out and finish him up with your hatchet.

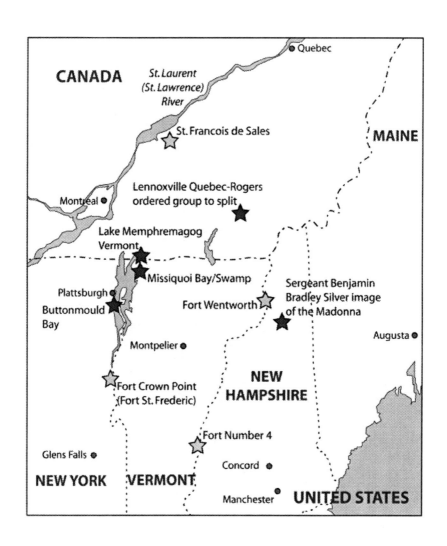

THREE

SURVIVAL

[O]ur options are reduced, until we have only one: to fight for survival.

—Morris K. Udall
U.S. Congressman, 1922–1998

DATE: *13 September to 9 November 1759*

WAR/CONFLICT: *French and Indian War*

LOCATION: *St. Francis River, north of Vermont, Quebec Province, Canada.*

MISSION: *To attack the Canadian St. Francis Indians (Abanakis) and "to chastise those savages with some severity."*

BACKGROUND

RANGERS WERE SCOUTING THROUGHOUT unexplored territory. Some were sent along the La Barbue, Upper Hudson, and Otter rivers. Others explored Isle aux Noix. One scout pitted a Ranger whaleboat against three wind-assisted French warships in a dangerous race that was won by the Rangers when one of the Rangers, a former New England fisherman, recommended converting blankets into sails. Although a field expedient measure, the Rangers won the day as the French gave up on approaching the British occupied Fort Crown Point.

In early September, Baron Jeffrey Amherst, the British Army commander, detailed a hazardous mission for Rogers' Rangers. Brigadier General Gage, the British military commander-in-chief in North America, was advancing slowly along the St. Lawrence River to La Gallete. Amherst's mission was to have Lieutenant James Tute, leading eleven Rangers and Regulars, take a whaleboat up the Sable River as far as possible to determine Gage's current situation and to conduct a reconnaissance for the movement of Amherst's own forces. Once the Rangers arrived at La Gallet, they would begin to move overland to the St. Lawrence.

Having taken only twenty-five days' provisions, Tute realized on day seventeen that he would soon be running short and sent a sergeant and four men back for supplies. Arranging to link up with these five men later, Tute continued marching and reached the St. Lawrence River on 20 September—twenty-seven days out—a few miles below La Gallete. Though starving, Tute resolved to complete the mission. Sneaking forward to reconnoiter La Gallete and take a prisoner, the Rangers were forced to abandon their attempt when Corporal Cauley of Gage's 80th Infantry Regiment deserted to the enemy—no doubt because of hunger.

Tute scribbled a note describing their desperate situation and dispatched a messenger to General Gage. As fast as its weakened condition would allow, the scout party then withdrew to its boat rejoining the five men left behind to secure provisions on the Sable River. Unfortunately, Tute's messenger was captured, the party's route of march revealed, and subsequently all were taken captive on the 22nd of September. After their capture, the Rangers were taken to La Corne at La Gallete and then later moved to Montreal.

During Tute's absence, Rogers forwarded a request to Baron Amherst to attack the Canadian St. Francis Indians. The St. Francis Indians had allied themselves with the French Canadians over the years and had participated in the killing of captured Rangers in the past. British tolerance for the tribe was finally exceeded when they learned that the Abanakis had attacked and captured one of Rogers' Ranger company commanders, Captain Jacob Naunauphtaunk, while he was under a non combatant "flag of truce." On 10 September 1759, Amherst approved Rogers' request and ordered him "to chastise those savages with some severity."

For the next two days, Rogers personally selected two hundred Rangers, Regulars, and Provincials who would accompany him on the mission. With the word out that Rogers was leading an expedition, destination unknown, Rogers was able to select the very best from the hordes of volunteers.

THE LEGACY

Rogers and his expedition departed in seventeen whaleboats on the night of September 13th with a total of 190 men. Reaching Buttonmould Bay, they remained concealed throughout the day and continued to the Otter River the following evening. While waiting for a dark evening so they could pass the three French warships guarding the mouth of the river, Rogers' force began to diminish in size as a result of injuries and personal illness inflicted by the harsh and unyielding forest environment surrounding them. In all, Rogers had to send back to Fort Crown Point forty-eight of his detachment, twenty-five Rangers and Stockbridge Indians [allied Indians from Massachusetts serving with the Rangers], sixteen Provincials, and seven Regulars. Despite losing 25 percent of his force, Rogers took comfort knowing that the remaining 142 officers and men were of "phenomenal endurance."

Days later, the warships finally sailed farther up the river one evening toward Fort Crown Point, allowing Rogers and his men to cross behind them in the darkness and proceed on their journey. Knowing that the Rangers traveled at night, the French had left some boobytrapped boats anchored at various points in hopes of ensnaring those inquisitive enough to look, but the Rangers, based on earlier experiences, evaded the ruse.

Arriving at Missisquoi Bay on 23 September, Rogers disembarked, concealing the boats in a tiny southern arm of the bay. Two days after they began their northward march, however, the two Stockbridge Indians who had been left with the boats for security caught up to the group and informed Rogers that the boats had been discovered by an enemy scout and burned.

Knowing full well that an enemy party would soon be in pursuit, Rogers decided to implement part of an old plan he had developed for a similar raid in 1756. Returning by way of Lake Memphremagog (a 30-mile-long lake located in south Quebec Province and northern Vermont) and the Connecticut River, Rogers dispatched Lieutenant McMullen, who was lame, and six other men to return to Fort Crown Point on foot. They were to cover the 120 miles in nine days and to request that Amherst send Lieutenant Samuel Stevens to the intersection of the Connecticut and Wells Rivers with provisions for Rogers and his men.

The French commandant at Isle aux Noix, Bourlamaque, posted 360 men at the burned whaleboat site in ambush and launched an additional three hundred men in pursuit of Rogers. Following the departure of the lame McMullen, Rogers and the remainder of his men marched for nine days, through the foot-deep muck of the Missisquoi swamps. Forced to sleep in hammocks strung between spruce trees above the bog, Rogers was able to evade the enemy, who was unable to track his party through the swamps. In an attempt to anticipate Rogers' moves, the French assumed he was moving against the closest Indian settlement, the Wigwam Martinique on the Yamaska River. Consequently, the French positioned 215 men to reinforce the Wigwam Martinique and another three hundred French and Indians at the mouth of the St. Francis River.

Rogers and his raiders reached the St. Francis River, fifteen miles north of the St. Francis village, twenty-two days after departing Fort Crown Point. Finding himself on the opposite side of the river and with no time to build rafts, Rogers led a group of his tallest Rangers across the river. With their heads barely out of the water, they were able to form a human chain by which the remainder—and shorter Rangers—could ford the river by pulling themselves along the bodies of the taller men. By the end of the day, the force was in position only three miles from its objective.

A reconnoiter that evening by Rogers and three other men found the Indians drunk in celebration of a wedding. One of the men, Lieutenant Turner, was captured when he moved forward for a closer look. Another of the group, a Stockbridge Indian private named Samadagwas, deserted to inform the town of the impending attack. A large number of the Abanakis Indians heeded the deserter's warning and abandoned the village to conceal themselves in the surrounding woods.

Rogers returned to the main body not knowing that Turner had been captured and that Samadagwas' warning had allowed for some of the enemy to escape. Moving into position at 3 AM, Rogers attacked the village from three sides a half hour before dawn. Despite the prisoner and deserter, and the departure of a large number of men, women, and children from the village, the surprise, incredibly, was complete. The remaining Abanakis were unable to organize any form of resistance and were quickly overwhelmed.

Many of the Indians refused to surrender, choosing to remain hidden in the cellars and attics of their well-built homes. Most of them would perish when the town was torched. All told, the Indian losses were placed between 65 and 140 Indians dead, and one town destroyed. Of the twenty women and children captured, all were released with the exception of Chief Gill's wife, Marie-Jeanne, and her five children, who returned with Rogers. Rogers' losses were one badly wounded—Captain Amos Ogden—and six lightly wounded. Lieutenant Turner's release as a prisoner was secured during the attack. To a degree, justice was served when the Abanakis killed Samadagwas, most likely in revenge for the deaths of their people and the destruction of their village.

With the raid over by 7 AM, Rogers had to make a hasty retreat south. Seizing enough corn from three of the village warehouses to sustain it for the first eight days of its march, the Ranger force withdrew, moving along the shore lines of Lake Memphremagog and the Connecticut River. Later, with provisions running out and pressure from his officers to do so, Rogers divided his force into nine groups, each with knowledgeable guides, to increase the force's ability to forage and hunt scarce game. At the appointed rendezvous, they would hopefully meet with Lieutenant Stevens and his provisions.

One of the groups was led by Sergeant Benjamin Bradley. Unknown to Rogers, he and a few of his men had pilfered the village and its church of many valuable possessions: a rare ruby, a golden calf or lamb, a great number of necklaces, silver broaches, sacred chalices, a solid silver image of the Madonna that weighed ten pounds, and even scalps. Weighed down with their ill-gotten gains, Bradley and three others struck out for his home in Concord, New Hampshire. Mistaking the Upper Cohase for the Lower Cohase, Bradley led his greedy followers erroneously into the White Mountains. For days, they wandered the mountains, trying to find their way out, only to have each succumb to the elements, some to be found the following year, others to remain missing forever—along with the silver Madonna.

Not far behind the Rangers, the Indians who had earlier fled the village were in hot and frenzied pursuit. Ensign Charles P. Avery's party was attacked and seven were taken prisoners. Another party of twenty men under the leadership of Ranger Lieutenant Turner and Regular Lieutenant Dunbar were attacked. Overwhelmed, the two officers and ten others were killed. The remainder escaped and was later rescued from starvation by Rogers. Fortunately, these were the only two detachments to be attacked.

For the remainder of the groups, the worst situations required exceptionally desperate measures—"survival of the fittest" in its purest form. Survival reached the depths of par-boiling leather accoutrements for food, while others turned to boiling birch bark. A party led by Sergeant Evans was so weakened by hunger that, coming upon the massacred remains of Dunbar's and Turner's men, they were reduced to cannibalism and "most of them sliced off choice portions." Taking some of the cadaver meat with them, the group moved on. For two days, Evans refused to eat the human flesh—unknown is the reason why. Was it rotting and putrid or was he revolted by the thought? Finally giving in, he cut a piece from one of three human heads he found in one of his Rangers' large knapsacks. Broiled in coals, he ate the piece. While declaring it the "sweetest morsel he ever tasted," he promised he would rather die of hunger before doing it again.

Evans' group was not the only one resorting to cannibalism. Lieutenant George Campell's party also stumbled upon the mutilated and butchered Rangers after Evans. They gorged themselves on the remains, with some of the Rangers consuming the flesh raw. Lieutenant Phillips' group of sixteen men marched directly for Fort Crown Point via the mouth of the Otter River. Nearly incapacitated by hunger, they were about ready to sacrifice and eat one of their three Indian prisoners when a Ranger was lucky enough to kill a muskrat. Divided among all the members of the party, it proved to be just enough to keep them from becoming cannibals and to get them to the mouth of Otter River where they were met on 8 November by fifteen Rangers in three whaleboats loaded with 150 pounds of biscuits and a gallon of rum "to refresh the starving Rangers." Another group was able to barely feed itself with the carcass of an owl.

Rogers' own small party sustained itself by sheer willpower. Unfortunately, Lieutenant Stevens, sent with provisions, failed to meet Rogers' party at the rendezvous point. Immediately upon arriving at the point, Rogers was stunned to find the still warm embers of Stevens' fire, the Ranger having just departed hours before to establish a new unauthorized base of operations five miles farther south. Rogers' men were able to hear Stevens' signal guns farther down the river but the lieutenant became alarmed when he heard the discharge of Rogers' weapons in return. Believing Rogers and his men to be Indians, Stevens repacked the provisions in his canoes and returned to Fort Crown Point on 30 October, leaving Rogers and his exhausted, emaciated men behind.

It would be ten more days before Rogers' arrival at Fort Number Four. Canoes loaded with provisions for his starving men were immediately dispatched to provide sustenance to the starving Rangers scattered about. Some were sent to Fort Wentworth while other similarly loaded canoes embarked up the Merrimac River. For thirteen days, seventeen men paddled the rivers. Of the 138 officers and men, including Rogers, who crossed the river to attack St. Francis, eighty-seven returned; a loss of 37 percent of the raiding force. Of the forty-nine men to die following the raid on St. Francis, only seventeen were combat-related deaths. All the rest died of starvation. Considering that 65 percent of the casualties could be attributed to the culpability of just one man, Stevens, and that the raid covered 400 miles taking nearly sixty days, the feat was an extraordinary military achievement. With this raid, Rogers was at the pinnacle of his career.

OBSERVATION

Food, especially during prolonged movement through vast, undeveloped territory, has always been of significant concern for military forces throughout history. As Napoleon once observed, an army marches on its stomach. It is a major problem that is further compounded when a significant portion of the terrain is swamp-like and the distances involved demanded constant movement—for swamps provide little in the way of plant nourishment and continuous movement virtually eliminates the ability to hunt wildlife.

Numerous campaigns in history have failed for want of provisions. That Rogers was able to complete his mission and return is a tribute to the field craft, fortitude, and willpower of his men.

For nearly two months, Rogers and his men subsisted off what they could carry on their backs or forage off the land. And, for those few desperates on the verge of starvation, they had no choice—cannibalism.

Fortunately, nutritional deprivation of today's modern Rangers is obsolete. Operations are always planned with 'Class I'—the Army's term for food—in mind. Even those who parachute into combat with only the equipment and supplies they carry on their backs do so with two to three days of meals—MREs: Meals, Ready to Eat—in their rucksacks. A single MRE is a dehydrated and heavily condensed food source that contains twice the normal daily adult requirement of calories—approximately 5,000 calories vs. 2,500.

Though there are certainly times when Rangers can, and do, go hungry during combat operations, our development of these dehydrated MRE rations, concentrates, vitamins, and power bars and the ability to carry or deliver such condensed sustenance ensures that today's soldiers never go without life-sustaining nutrients for too long.

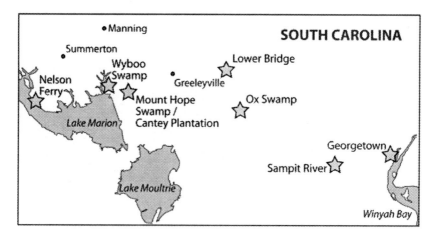

FOUR

INTREPIDITY

To fight the enemy bravely with a prospect of victory is nothing; but to fight with intrepidity under the constant impression of defeat, and inspire irregular troops to do it, is a talent peculiar to yourself.

Nathanael Greene
American Revolutionary War Continental Army Major General, 1742–1786, to Francis "Swamp Fox" Marion, Partisan and Continental Army Brigadier General, 1732–1795.

DATE: *5–28 March 1781*

WAR/CONFLICT:The *American Revolution*

LOCATION: *Vicinity of Georgetown, South Carolina*

MISSION: *Maneuver, engage, and defeat the English "Buff's Regiment"—one of the finest combat units in the world with an American force of irregulars.*

BACKGROUND

THOUGH TITLED RANGERS BY George Washington and acknowledged by a grateful American people, Morgan's Rangers were not, essentially, Rangers in the sense of being the uniquely trained, motivated, and challenged soldiers. With such traits in mind, there was only one group of American revolutionaries who could be truly considered the finest example of early American Ranger prowess. This highly successful group of partisans—guerrillas—was known as "Marion's Brigade."

Formed in 1780 to counter the brutality of British General Cornwallis, the brigade was commanded by Brigadier General Francis Marion, one of the boldest and most dashing figures of the American Revolution. Taking refuge in forests and swamps, for which he earned the moniker the "Swamp Fox" from his enemies, Marion and his brigade kept the British off-balance with rapid movements that captured British troops, destroyed supplies, and disrupted lines of communications.

Between 15 August 1780 and 8 September 1781, Marion and his men baffled the British in South Carolina. Hiding in his lair on Peyre's Plantation or Snow's Island by day, and stealthily emerging to strike around midnight, Marion and his men would wreak havoc on British units from White Marsh to Black Mingo before fading once again into the morasses of the Santee or Peedee Rivers. Immortalized in song and story, Francis Marion became a hero of the Revolution second only to the Continental Army Commander George Washington.

Having fought with the Continental Army regiments of South Carolina since the start of the Revolution, Colonel Francis Marion found himself the most senior and respected officer in that state by the summer of 1780. Unfortunately, much of the Continental Army in that southern state had been mauled in the interim and Marion was reduced to being a commander in name only. Aware that there was a group of friendly Scot-Irish militia in Williamsburg already armed and under the command of Major William James, Marion proposed that he assume command of this unit and conduct a boat-burning raid up the Santee, south of Camden. Approval was obtained from the Continental Army commander in the south, General Gates, and on 15 August Marion and a small group of South Carolina officers and men rode out of the American army encampment and into American history.

Marion and his men arrived at the Williamsburg militia camp at Witherspoon's Ferry late afternoon of 17 August. Though short in stature and limping on a bad leg, Marion, 48, was met by an exhilarated Major James and his militiamen. Without hesitation, the militia readily accepted him as their commander, though he had no legal authority to command as an officer of the Continental Line without rank in the militia. Soldiers of the Continental Line were, relatively speaking, active duty men who served in the Army while militia could be considered analogous to today's National Guard. Though each is in its own right a military organization of the United States, each is limited in authority over the other.

Quickly getting to know his men, Marion wasted no time carrying out orders from Gates by moving along the Santee in the proximity of Lenud, burning flatboats and destroying canoes as he moved north toward Camden.

For the Continentals, the situation in South Carolina had reached its lowest point. All major military units within the state had been routed or captured following a disastrous battle in the vicinity of Camden on 16 August between General Gates and Lord Cornwallis. The British occupied the state capital; the state government was in disarray. South Carolina lay defenseless before the Crown. Despite the debacles, Marion proved to be at his best when the odds seemed worst. Concealing the Camden disaster from his troops, he continued his mission of marching and burning, operating as an independent force.

On 26 December, Lieutenant Colonel John Watson and His Majesty's 3rd Regiment of Guards marched to an area ten miles north of Nelson's Ferry where they constructed a formidable fortification on Wright's Bluff overlooking Scott's Lake. This fort was complete with obstacles, platforms, and gun ports for his two cannons. An eighty-foot Indian burial mound was in the center of the fortification on which he placed a firing position for his sharpshooters. Upon completion, the regimental commander vainly named it after himself, Fort Watson. Unknown to Lieutenant Colonel Watson, Fort Watson was in Francis Marion's sight.

THE LEGACY

Because Marion vexed the British thoroughly, the decision was made by Cornwallis' replacement, field commander Lord Francis Rawdon, to try and eradicate this pest once and for all. A double-pronged attack was planned with Lieutenant Colonel Watson marching down the Santee to attack Marion directly while Colonel Welbore Ellis Doyle crossed the Lynches River to cut off the rebel's escape. On 5 March, Watson began his movement and marched from Fort Watson, encamping south of Nelson's Ferry. That same evening, one of Marion's scouts reconnoitered the camp and rode off to warn his brigadier.

Marion was located past Murry's Ferry. When he learned of Watson's deployment, he deduced that this was a strategic attempt by the British to drive him from the southern region of the state. Watson's Buffs was one of the exemplary regiments in the British army. Marion knew that, as with all previous battles, he had to rely on his own wits, and facing off with Watson would be a supreme test.

Undaunted, the Swamp Fox ordered an immediate advance and moved to position himself at Wiboo Swamp. There, Marion waited, knowing that Watson would eventually appear. Soon, the British regiment marched into view. Both Watson and Marion rode out to face each other across a quarter-mile causeway spanning the mire and marsh of the swamp.

Watson did not take long to initiate the fight. Sending forth his Loyalist cavalry to cross the causeway, Marion met the charge with one of his own led by Colonel Peter Horry and his mounted men. Following a brief skirmish on the narrow land bridge, both sides withdrew. Advancing his main force as support, Marion again ordered Horry to charge. Undaunted, the British Regulars held their ground, employed their field cannon, and repelled the attack.

As Horry withdrew, Watson's mounted Tory dragoons—mounted infantry—advanced right behind, successfully crossing the causeway. A countercharge of the remainder of Marion's cavalry pushed the dragoons back across the land bridge. With his cavalry defeated, Watson advanced his guards. Imposing and silent, highly disciplined and trained, impressive in their bright red uniforms, the guards led with bayonets that gleamed and sparkled on the ends of upraised muskets. Realizing that he had done enough and that his men would not be able to stand in the open ground against such a disciplined and highly trained force, Marion had his men remount and follow him, leaving that field of battle to Watson.

The morning following their initial engagement, Watson resumed his march down the Santee Road with Marion slowly backing away before him, remaining out of the range of the British artillery. That evening both sides encamped, though Marion did order his night patrols to take shots at Watson's sentinels.

Morning arrived and Watson again resumed his march with Marion leading the withdrawal. At Mount Hope Swamp, Marion had his men disassemble the bridge. A covering force of Marion's riflemen was swept aside by Watson's artillery, allowing Watson's guard to ford the stream unopposed. The cat and mouse advance continued on the far bank.

Crossing the road that led from Murry's Ferry to Kingstree, Watson continued to follow Marion toward Georgetown. Soon though, just as Marion had expected, the British guardsmen wheeled around and made their way back to the Kingstree intersection. Lower Bridge was only twelve miles away. To cross it would put Watson deep within the heartland of the Whig resistance. (Note: Whig's were Americans loyal to the Crown; Tories were Americans in rebellion, such as Marion, against the Crown.) This was the decisive point of the campaign that required Marion to attack.

Seventy of Marion's men, thirty of whom were sharpshooters, were dispatched to ride ahead to secure and hold the Lower Bridge under Major James. With knowledge of the land, James and his men were able to secure the bridge before the arrival of Watson's dragoons. Within minutes, the planks had been removed from the center of the bridge and the stringers burned on the east end. The sharpshooters were placed at the abutment where they had the clearest shot at the far end of the bridge while the remaining musketeers secured their flanks. Marion and his brigade soon arrived and forded the river. Pleased with James' deployment, Marion reinforced him with an additional company and moved to a reserve position to the rear and out of sight.

The east bank of the river was low, open, and swampy. The west bank was a high bluff with the roadway passing down through a ravine to the bridge. The distance from bank to bank was the perfect sharpshooter distance: fifty yards.

Realizing that he needed to clear the far bank first before his soldiers could safely enter the defile to cross the bridge, Watson emplaced his cannon to cover the movement. However, the bluff was too high, and the artillery canister fire just passed over the heads of the defenders on the east bank. An attempt to move the cannon and depress the muzzles led to the crews being run off by the highly accurate fire of the sharpshooters.

Curtailing his attempt to clear the far side first, Watson formed his men in column and ordered the first column forward with its captain out front. Upon reaching the ford site, the captain was killed outright with one shot from Marion's sharpshooter commander. Four men attempting to recover his body were also killed in sequence.

Baffled about what to do next, Watson waited on the bluff until evening, then withdrew to a plantation a mile north of the bridge. To the plantation's Whig owner, Watson acknowledged he had "never seen such shooting before in his life." While Watson established his headquarters in the plantation house, Marion and his men bivouacked on the wooded ridge south of the ford. Marion and his men had won the day's skirmish. But they were engaged with one of the finest regiments in the British army and thus, in the world. Marion knew that if he did not continue to press, they would return—with a vengeance.

Before daybreak, Marion roused his men and deployed his troops. To get things off to a spirited start, he dispatched a detachment of sharpshooters across the river to the plantation where Watson and his troops were housed with orders to shoot his sentries and to wreak havoc. These sharpshooters were proving to be so successful that Watson felt compelled by noon to redeploy his regiment to a large open field about a half-mile away, believing this would minimize his losses.

Unfortunately for Watson, the lack of trees and concealment did not diminish the fires or their accuracy. With his men in a panic and the number of wounded and suffering growing, Watson shelved his pride and addressed a letter to Marion requesting permission to send seven of his most seriously wounded through the lines but also rebuking Marion for conducting himself in a manner contrary to civilized war. Marion's response was to once again reiterate his position that he was only responding in kind to earlier British transgressions. He did, however, send a pass for the safe passage of the wounded and their attendants.

Watson and his men remained where they were, subsisting off the plantation and commandeering anything else they needed from those homes around them. Scouts abounded about the British regiment. Watson was cut off from outside information, for his messengers could not break through Marion's cordon. As his situation grew more desperate with each passing day, Watson finally decided to retreat.

Leaving his dead in an abandoned rock quarry and loading his wounded on wagons, Watson and his crack regiment began their withdrawal at the double-time on 28 March toward Georgetown. Seven miles later at Ox Swamp, he encountered destroyed bridges and abatis—downed trees—across the causeway. The situation was not good. Swamp to his left and right, passage blocked to the front and Marion closing in from the rear, Watson had to choose quickly whether to flee or fight.

Choosing escape, Watson abruptly wheeled his regiment to the right and proceeded to move at a quick pace across fifteen miles of marsh and pine lands to the Santee Road. In pursuit was Peter Horry, firing on the retreating column from every bush and thicket. The infantry moved at a trot the entire way, stopping only to fire a volley to their rear. Within the formation were the wagons, periodically stopping to gather more wounded—or dead.

Nine miles from Georgetown, the harried regiment approached the Sampit River. Having dashed ahead to destroy the bridge, Horry and his men were positioned fifty yards across the river. Undaunted by the destroyed bridge or the rebels' fire, the lead guardsmen closed column and plunged across the river. As the advance guard made its way across the river, Marion fell upon the rear guard.

A quick fight ensued with heavy firing. Watson rode back to rally his men only to have his horse shot from under him. Mounting a second animal, he ordered his artillery to open fire with canister. As Marion's men turned back from the fires, Watson loaded two wagons with wounded. Leaving twenty dead from the engagement, he, the wagons, and the remainder of his regiment forced the ford.

Safely camped in the vicinity of Georgetown later that evening, Watson was extremely bitter, complaining that Marion and his men would "not sleep and fight like gentlemen." Instead, "like savages," they were "eternally firing and whooping around us by night, and by day waylaying and popping at us from behind every tree!"

OBSERVATION

Marion's battle with Watson is a classic, textbook example of inspired and resolute courage. Overall, Watson had made the most critical mistake that a commander could make—he had underestimated the capabilities of his enemy. Even worse, he had underestimated the capability and ability of the commander he faced, Francis Marion.

Marion, on the other hand, had not underestimated his enemy. He knew what he faced and he had devised an unconventional plan of action to defeat a conventional force. In the words of General Greene, Marion had inspired his men "to fight with intrepidity." As a result, the Swamp Fox and his partisan group of "uncivilized" American militia had amazingly and soundly defeated the Buff's Regiment, one of the finest fighting combat units in the world.

These series of engagements also illustrate the oft unrecognized power of guerrilla forces over conventional forces. Relentlessly harrying and standing to fight only from an advantage, Marion's Brigade eviscerated the combat effectiveness and will of Watson's regiment. Ruefully, the French experienced the same style of warfare and ultimate defeat at the hands of the Viet-Minh during the War in Indochina.

In an interesting bit of irony, one of Marion's decedents would later serve as an American armor officer in the Indochina War, fighting against the Viet-Minh who were waging the same kind of war his great ancestor had fought nearly two centuries before.

Brigadier General Francis Marion

History will record his worth, and rising generations embalm his memory, as one of the most distinguished Patriots and Heroes of the American Revolution.

Francis Marion's tombstone

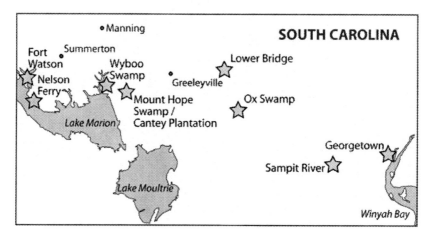

FIVE

INITIATIVE

Some officers require urging, others require suggestions, very few have to be restrained.

George S. Patton, Jr.
Second World War U.S. Army General, 1885–1945

DATE: *16–23 April 1781*

WAR/CONFLICT: *The American Revolutionary War*

LOCATION: *Santee River, South Carolina*

MISSION: *Lay siege to and capture Fort Watson.*

BACKGROUND

UNFORTUNATELY FOR MARION, HE was unable to bask in his incredible victory against Lieutenant Colonel Watson and the Buffs' Regiment for long. His base of operations at Snow's Island had been located and destroyed during his absence. Marion's focus on Watson had led him to neglect putting out scouts or placing guards at various ferries. Consequently, he was surprised to learn from a messenger the day after his victory that an enemy force under the command of Colonel Welbore Doyle and composed of his New York Volunteers had successfully made its way through the swamps and bogs towards Snow's Island, following the trampled paths and trails cut out by Marion's riders.

Though Marion's deputy, Colonel Hugh Ervin, and the small stay-behind contingent attempted to put up a spirited defense, they were quickly overwhelmed by the superior force at the boat landing on Clark's Creek. Withdrawing to the island's cottage after having suffered seven dead at the boat landing, Ervin was forced to release the twenty-six prisoners located on the island, in addition to leaving behind an additional fifteen American militia who were too ill to escape.

Unable to bring any of their stores or equipment with them, Ervin and his survivors abandoned Snow's Island. Doyle, having captured the island without losing a single man, torched the structures and supplies, and hastily recrossed Clark's Creek concerned that he would be caught in a trap should Marion and his men suddenly return.

Marion and his exhausted men pushed hard to reach the island prior to Doyle's departure. The nature of the militia was showing again as the men began to slip off without leave or authority as the force made its way through the countryside. By the time Marion arrived at Indiantown, the proud brigade that had just defeated one of the world's finest regiments had been reduced to seventy men without a single shot being fired at it.

Refusing to become despondent, Marion launched his scouts at daybreak to locate Doyle. They returned with reports that placed Doyle's men looking for food at British sympathizers' plantations south of the Lynches River. Hugh Horry and his mounted infantry were dispatched to drive them off. Finding Doyle's men ransacking a plantation, Horry and his men killed nine and took sixteen prisoners. The fleeing raiders were pursued back to Doyle's camp across the Lynches River.

Arriving just as Doyle's panicked rearguard was scuttling the ferry, Horry had his sharpshooters take up position and begin to place effective fire on Doyle's camp. Doyle responded aggressively by forming his infantry in ranks along the river and returning volley fire. But his fires proved to be ineffective at that range, especially against the well-covered and concealed Americans. Realizing he was in a no-win situation, Doyle struck camp and marched off toward the Peedee River.

Despite their great victory against Watson, Marion and his men were starting to become discouraged. Marion had now been in command for seven months. Their recent deployment against Watson had them in the field for three continuous weeks. Some of the militia had been killed, with many others wounded. The vengeful British had burned over one hundred of their homes to the ground. What little supplies they'd accumulated had been destroyed on Snow's Island. And no matter how many victories they achieved, it seemed as though the British always returned stronger the next time.

The remaining men were depressed. Finally, after giving it much thought, Marion had his small force mustered. Standing before them in a dirty, faded, and worn Continental uniform, looking gaunt, shaggy, and unshaven, Marion addressed the remnants of his brigade, reminding them of their duty, their service, and the sacrifices they'd already made. He reminded them of the taxes demanded by a government that truly did not represent them. He reminded them of the outrages perpetrated in the name of the King by his soldiers and allies alike. In closing, he left them with this one final thought.

> Now, my brave brethren in arms, is there a man among you who can bear the thought of living to see his dear country and friends in so degraded and wretched a state as this? If there be, then let that man leave me and retire to his home—I ask not his aid. But, thanks to God, I have now no fears about you. Judging by your looks, I feel that there is no such man among us. For my own part, I look upon such a state of things as a thousand times worse than death. And as God is my judge this day, that if I could die a thousand deaths, most gladly would I die them all rather than live to see my dear country in such a state of degradation and wretchedness.

To cheers and rousing enthusiasm, the Swamp Fox had revived the lagging spirits of his men. In response to their shouts that they would fight beside him to their death, Marion declared, "Well, now, Colonel Doyle, look sharp, for you shall presently feel the edge of our swords!"

THE LEGACY

Calling in all his riders, Marion struck camp and led his men upstream of the swollen and flooded Lynches River to a swamp behind the James Plantation. With no illumination, Marion spurred his mount, Ball, into the dark waters. For many, such as Peter Horry, no other experience had ever placed them "so near the other world." Horses, men, and equipment were swept away. Amazingly, no one was killed but many did have to wait until daylight to continue their movement across the river, having found themselves caught in trees or sandbars in the raging water.

Fires were built on the far shore to thaw bodies and to dry clothes. Then, weapons primed with dry powder, Marion and his men rode on. In the evening, they arrived at the plantation of a noted Whig named Burch to find that Doyle and his men had encamped there just the day before. Burch informed Marion that a message from Lord Francis Rawdon, Cornwallis' replacement as field commander in South Carolina,was delivered that evening, after which Doyle destroyed his heavy baggage and hastily set out toward Camden, sixty miles away. Marion would soon learn that General Nathanael Greene, who had relieved Gates as the commander of the Continental Army of the South, had elected to return with his army to South Carolina in late March.

Though disappointed that his efforts had not led to Doyle's destruction, Marion was exhilarated by the fact that the Continental Army's focus was once again South Carolina. Intuitively, Marion visualized another strategic plan as a result of these changing sets of circumstances. Directing Lieutenant Colonel Henry "Light Horse Harry" Lee—the father of the future Confederate commander, Robert E. Lee—and his legion, which had set out to rendezvous with him on 4 April, Marion and his brigade deployed to intercept Watson, whom he knew would be marching from Georgetown toward the British concentration in Camden. Moving through Williamsburg, though, Marion was once again faced with the continuing problem of homesick men, who were not regular army, deserting as they passed through their home region. The brigade's strength, originally one hundred or more, was down to eight men by the time it reached the bridge over the Black River.

Lee and his men effectively linked up with Marion and his severely depleted brigade on 14 April. The affection and respect of each commander for the other was genuine and, after much discussion of what had transpired since their last meeting, they both began to coordinate their joint campaign. Lee's knowledge of the plan Greene had for the Carolinas figured immensely in their strategy. At daybreak on 16 April, they began to implement their plan.

With Lee detaching a patrol of dragoons—mounted infantry—toward Georgetown to report on Watson's movements, the remainder of the force moved through Nelson's Ferry to Wright's Bluff to begin their siege of Fort Watson. The fort's commandant, Lieutenant James McKay, assessed his situation. Morale was high and quantities of ammunition and food plentiful. And, though Marion's sharpshooters cut him off from his water sources, he was able to have a well dug within the confines of his stockade, striking water on 18 April. Despite the fact that Watson had taken the only cannon with him on his deployment, McKay felt confident that he had enough resources to hold the redoubt (fort).

Realizing that the siege could become somewhat lengthy, Lee developed an alternative and requested in writing that Greene loan the brigade a field piece from his regular army Continental forces. "Five minutes will finish the business, and it can be immediately returned." Unfortunately, disaster in the form of smallpox struck Marion's men as they awaited the cannon. Many of the militia had never been exposed to the illness and thus had never developed any immunization antibodies. The infectious disease spread rapidly throughout the camp. The men deserted in large numbers. The healthy fled to avoid coming down with the affliction; the ill departed to seek medical care and nursing.

Having been chastised earlier on the 16th by his oldest and most respected friend in the army, General Moultrie, for supposed transgressions against the local populace by his militia, Marion was already feeling a bit dejected. The desertion of his soldiers and lack of success just compounded the problem further. Lee grew concerned over Marion's growing despair, for he was thoroughly sympathetic to the Carolinian's situation, knowing full well that it was only through the Swamp Fox's efforts that the British had not totally overrun South Carolina after the colonial army's retreat.

Attempting to seek some form of acknowledgment for Marion's previous services, Lee wrote Greene, requesting that he write a long letter to his militia brigadier. Though Greene held Marion in high regard, he did not feel the same about the militia. While there were "brave and good officers" in command, "the people with them just come and go as they please." While words of appreciation or thanks in regard to the militia would not be coming from Greene at the moment, he soon would have reason to do so. Though Greene would not send the letter per Lee's request, he did send a cannon with gun crew and a detachment of infantry to escort it.

Unaware that their request for the cannon had been approved and that it was on the way, Marion, given his depressed state, began to seriously consider lifting the siege and withdrawing. To date, his force had suffered two killed and six wounded with nothing to show for its efforts.

Fortunately for the militia, there was an alternative to the cannon and a newly assigned officer in the Brigade, Major Hezekiah Maham, proposed it. Considering and then accepting Maham's recommendation, Marion set his men to work constructing an old-style siege tower built of logs. Traversing the countryside, Marion's men located, chopped down, stripped, and shaped trees into the lumber necessary to construct the siege weapon. The design called for a floor in the tower that was higher than the fort walls before them. To protect the marksmen who would be inside, the front of the tower was reinforced with a dense shield of timber. Finally, during the early morning hours and under the cover of darkness, the mobile fortification was pulled forward into position. Though the defenders could hear movement in the blackness of the night, they had no idea of what would soon be before them.

The morning of 23 April dawned to display a tower filled with sharpshooters looming over McKay and his men. The piercing noise of the sharpshooters' bullets echoed throughout the fort while the return fire of the British muskets failed to penetrate the tower's thick shield. Unable to raise their heads above the parapet—wood walls—unless they wanted to be picked off, McKay could only watch in frustration as militia pioneers breached the palisades—sharpened wood stakes pointed towards the enemy like bayonets. that encircled the fort and began to attack the logs of the stockade's wall itself. Lee's infantry with fixed bayonets stood prepared, behind the pioneers. A white flag was soon raised from inside the fort. The eight-day siege was about to come to an end.

The American commanders offered generous terms to McKay and his men as recognition of their bravery. Officers were allowed to sign paroles (agreeing to not take up arms again in exchange for their freedom), keep their swords and baggage, and move to Charleston to await an exchange while both the Regular and Loyalist troops were treated alike as prisoners of war.

The fall of Fort Watson established an operation that others would emulate along the Santee River. American troops for the first time had toppled a British fortress in South Carolina. Even better, it had been accomplished by a joint militia and Regular army initiative. In Marion's official report, he gave great credit to Maham and Lee.

Departing Fort Watson to encamp on Bloom Hill in the High Hills, Marion arrived to find the long letter Lee had suggested Greene write.

> When I consider how much you have done and suffered, and under what disadvantage you have maintained your ground, I am at a loss which to admire most, your courage and fortitude, or your address and management. Certain it is no man has a better claim to the public thanks, or is more generally admired than you. History affords no instance wherein an officer has kept possession of a country under so many disadvantages as you have; surrounded on every side with a superior force; hunted from every quarter with veteran troops, you have

found means to elude all their attempts, and to keep alive the expiring hopes of an oppressed Militia, when all succor seemed to be cut off. To fight the enemy bravely with a prospect of victory is nothing; but to fight with intrepidity under the constant impression of defeat, and inspire irregular troops to do it, is a talent peculiar to yourself. Nothing will give me greater pleasure, than to do justice to your merit, and I shall miss no opportunity of declaring to Congress, the Commander-in-chief of the American Army, and to the world in general, the great sense I have of your merit and services.

It would seem that General Nathanael Greene's opinion of Brigadier Francis Marion's contribution and the abilities of his militia had risen greatly with the fall of Fort Watson.

———

Buoyed by his success against Fort Watson and by his commander's letter, Marion continued to carry the war against the British. A detachment of eighty men was dispatched to conduct a raid through the High Hills, along Rafting Creek, to the source of the Black River. Instilling fear within the British and disrupting the flow of supplies throughout the region, the raid resulted in significant pressure being placed on the local enemy garrisons.

The general success of the American Army and its proximity to his location, soon forced Rawdon to abandon Camden. Shortly thereafter, Greene met with Marion for the first time. Each grew to admire the other as their talk on strategy progressed. Following this meeting, Marion moved down the Santee as Henry Lee marched on Fort Grandy. Laying siege to the fort on 14 May, Lee was able to negotiate a surrender on 15 May of the 340-man garrison, including two cannons.

———

OBSERVATION

SEIZING THE INITIATIVE MEANS not only setting something in motion, it also means a readiness to embark on bold new ventures. Nothing can be bolder or newer to light infantry Rangers than laying siege to a heavily defended fort.

Rather than allow the enemy to seize the initiative by setting the time and place of battle with forays from behind stockade walls, Marion seized the initiative from the enemy with cunning by attacking what he believed was their strength. But laying siege, alone, was not the answer for time and disease are long drawn-out methods that could work against him. His officers devised an exceptionally unique and innovative plan to eventually breach the enemy's wall. It is relatively easy to assume that Major Mahan may have been well read on ancient warfare because his siege tower was a close replica to those used by Julius Caesar during his conduct of siege warfare in Gaul. Mahan's initiative in applying an ancient siege engine to modern warfare was indeed a tactical inspiration. In the end and in stark contrast to the modern weaponry of the day, Fort Watson fell before the initiative of a centuries old military tactic.

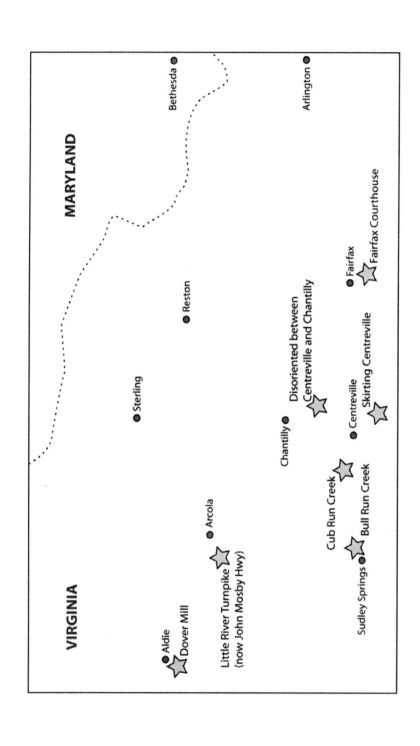

Six

BOLDNESS

There are occasions when daring and risky operations, boldly executed, can pay great dividends.

— **General Mathew B. Ridgeway**
(Second World War & Korean Conflict US Army General,
1895–1993)

DATE: *8–9 March 1863*

WAR/CONFLICT: *American Civil War*

LOCATION: *Fairfax Courthouse, Virginia*

MISSION: *Surreptitiously move through enemy lines under the cover of darkness, capture a senior Union officer, and return to friendly lines.*

BACKGROUND

THE CIVIL WAR—OR "WAR of Rebellion"—forced the South to conduct an unconventional and partisan style of warfare along the fringes of the Confederacy. These attacks were executed by a total of 428 units, officially and unofficially designated as "Rangers." The most successful of these units was Mosby's Rangers. Officially designated the 43rd Battalion of Virginia Cavalry, Mosby's partisan Rangers were lead by John Singleton Mosby, who believed that by applying aggressive action, he could compel his enemies to "guard a hundred points," thus expending valuable troops and resources needed elsewhere. These Rangers were particularly active in Virginia and Maryland from 1863 to 1865 and had an excellent reputation within the Confederate army.

In February 1864, the Confederacy, to maintain better control, re-organized the multitude of partisan and guerrilla-style forces that existed; two partisan commands, however, were exempted from the reorganization and allowed to retain their independence, John Singleton Mosby's command being one of them.

Mosby first voiced the concept of a Ranger-style cavalry unit in December 1862. Following a raid against Burnside's Army of the Potomac, Confederate Brigadier General Jeb Stuart and his cavalry withdrew to Loudoun County in northern Virginia for a few days of rest and rehabilitation. As they relaxed, Mosby—an enlisted member of his command—approached Stuart to discuss an idea that he'd been considering for quite some time. Mosby was convincing and Stuart's support of Mosby's idea on the irregular style of conducting war led to the start of twenty-eight months of attacks, ambushes, and raids against the Northern forces within the region.

The area of Mosby's operations was 100 miles long, running between the Federal capital of Washington, DC, and the Confederate capital of Richmond, Virginia. The region would ultimately encompass the counties south and west of Washington, south of the Potomac River, and those in the northern portion of the Shenandoah Valley at the base of the Blue Ridge Mountains. Much of this area was hills and wooded mountains interspersed with fertile farmland that was conducive to guerrilla warfare.

Based in southern Loudoun County and in the northern and western portion of Fauquier County, Mosby and his men would eventually strike eastward into Culpeper, Fairfax, and Prince William Counties. Operations would also take him westward into Warren, Clarke, Jefferson, and Frederick Counties. Much of this region would become known as "Mosby's Confederacy," for many of his men would be quartered with their parents while others would reside with family or friends. By Mosby's definition, his Confederacy encompassed the roughly 125 square miles that ran south along the

Blue Ridge Mountains from Snickersville to Linden at the Manassas Gap, east through Upper Fauquier to The Plains, north along the Bull Run Mountains to Aldie, and west back to Snickersville.

THE LEGACY

Mosby's philosophy was simple. "If you are going to fight, then be the attacker," and his initial successes resulted in the Union cavalry remaining in camp for weeks. This additional time allowed Mosby to plan and conduct what would prove to be possibly his most notable and audacious raid. Mosby issued an order to muster on 8 March 1863 at Dover Mill, just west of Aldie in Loudoun County. In groups of two and three, twenty-nine of Mosby's riders assembled on a miserable, rainy day.

With the men ready by dusk, Mosby seated on his horse announced, "I shall mount the stars tonight or sink lower than plummet ever sounded." The men followed blindly, not knowing that their objective was to penetrate Union lines to capture Colonel Percy Wyndham, the commander of a cavalry brigade, who was reported to be camped in his headquarters at Fairfax Court House. The mission was a vendetta for Mosby. Frustrated by Mosby's recent successes and his own command's failure to stop the Confederate Rangers' forays around his capital, Wyndham had resorted to calling Mosby "a common horse thief" and had threatened to burn the homes of those he believed supported the partisans. Taking Wyndham's comments rather personally, Mosby had decided to "put a stop to his talk by gobbling him up in bed and sending him off to Richmond."

Proceeding eastward on the Little River Turnpike, the Rangers moved in a cold rain through the pitch black of night and fields of mud. Leading the column between the flickering Union campfires to the Warrenton Turnpike, Lieutenant James F. "Big Yankee" Ames—a Union deserter from the 5th New York Cavalry—was able to avoid discovery by guards and turn northward toward Fairfax. A mile and a half from the courthouse, they veered away again, this time eastward to avoid the campfires of Vermont infantry troops. Unfortunately, in the process of their maneuvering, the column had split and nearly half of the Rangers had become disoriented in a pine forest between Centreville and Chantilly. It was 2 AM before the column had reunited and entered Fairfax.

Quietly moving through the town and capturing guard post sentinels, the raiding party quickly cut the telegraph wires, secured the horse stables, and occupied various other buildings. Arriving at the building he thought housed the Union cavalry commander, Mosby was informed that his information was incorrect and that Colonel Wyndham had been unexpectedly summoned to Washington, DC Disappointed, Mosby opted to make the best of it. An interrogation of some prisoners soon revealed, however, that an even greater prize, was theirs for the taking. The commander of the Vermont infantry brigade, Brigadier General Edwin H. Stoughton, was headquartered in town. The son of a wealthy family, Stoughton had been the youngest general in the army at the age of twenty-four. On this particular night, he had hosted a champagne party and had retired to bed to sleep off the effects of his revelries.

Mosby and some of his men walked to the brick home of Dr. William Presley Gunnell, where Stoughton was staying for the evening. A knock on the front door brought the head of one of the general's staff, Lieutenant Prentiss, out of a second-story window. Upon being informed that there was a "dispatch for General Stoughton," the lieutenant came down and opened the door.

Grabbing the startled Union officer, Mosby and his men forced their way into the house. Convincing Prentiss—with a bit of coercion—to lead them to his commander's bedroom, the Confederates found the young twenty-five-year-old general sound asleep. Accounts differ as to what happened next. Witnesses state that Mosby simply shook the general awake while Mosby's memoirs have him removing the covers, lifting the nightshirt, and unceremoniously spanking the Federal general's *derrière*!

In either case, the general found himself suddenly awaken and surrounded by a group of rude strangers. Confused and demanding to know who was present, the general was asked by the twenty-nine-year-old Confederate lieutenant who was wearing a captain's uniform: "General, did you ever hear of Mosby?"

"Yes, have you caught him?" queried the Union officer.

"No, but he has caught you," came the response. "Stuart's cavalry has possession of the Court House; be quick and dress."

Allowing the general to dress, though he was a bit slow about it, the group moved back to the town square with their captives by around 3:30 AM. With scores of dazed and confused Union soldiers scattering about the town trying to gain some semblance of order and trying to determine what was happening, Mosby and his men began to ride out of town with their thirty Union soldier captives and a large number of Union horses.

The column had only traveled a few hundred feet when the command, "Halt! The horses need rest!" was bellowed from a second-story window. Answered by silence, the voice continued, "I am commander of the cavalry here and this must be stopped." Realizing that he was addressing rebels below him, Colonel Robert Johnstone, commander of the 5th New York Cavalry, quickly retreated back into the house, as two of Mosby's men broke through the front door. Encountering Mrs. Johnstone in the hallway, the men were distracted just long enough to allow the buck-naked colonel to escape through the back door and unceremoniously conceal himself under an outhouse on the grounds.

Having lost the opportunity to capture Johnstone, Mosby and his men continued their withdrawal along the same route they had entered for approximately a half-mile. At that point, Mosby left the highway and began to move cross-country to throw off his pursuers, whom he knew would soon be following. The terrain was harsh and, with no moonlight, it was difficult to see. The column began to get strung out and prisoners began to escape, disappearing into the darkness. Finally reaching Warrenton Pike, Mosby turned over command of his column to William Hunter, who moved out at a fast trot with the group while Mosby and a second Ranger formed a rear guard.

Fleeing through the night, the raiders skirted the Federal camps located in Centreville and continued on. The clatter of their hooves, if heard, was probably mistaken as that of Union cavalry. Encountering the swollen Cub Run, the Rangers were only able with difficulty to get their horses across. As the new day's sun was rising, Mosby and a second Ranger led the raider force across the Bull Run creek and through that bloody field of battle of July 1861 and August 1862. Mosby's raid had netted one Union general, two captains, thirty enlisted men, and fifty-eight horses. Not a shot had been fired, not a man killed or wounded.

OBSERVATION

The "Stoughton Raid" was proclaimed by Confederate General Jeb Stuart as "a feat unparalleled in the war." As a result, Federal camps and headquarters began to feel insecure and vulnerable to the partisans' roving patrols. This acknowledgment of Mosby's prowess even made its way to the highest levels of Washington where a rumor circulated that the president and his cabinet might be the Confederate Ranger's next target. Acting on those fears, the Union army ensured that the link across the Potomac between Washington and Virginia, the Chain Bridge, was made impassable each night for a week with the removal of its flooring planks.

President Lincoln, though, was amused by the incident stating he "didn't mind the loss of the brigadier as much as the horses. For I can make a much better general in five minutes, but the horses cost one hundred and twenty-five dollars apiece."

All this angst the result of a Ranger's single bold raid.

Irregular warfare as practiced by the Mosby's Rangers and similar cavalry units had a greater impact on the Civil War than is generally acknowledged. The Union Army was forced to divert considerable infantry units to guard depots, armories, rail centers, main supply roads and rail lines as well as bridges to name just a few favorite targets. Moreover, Union cavalry and infantry forces were frequently sent afar to give chase to confederate raiding parties. Given that the Union, particularly the Army of the Potomac, was continually strapped with manpower shortages due to a poor recruitment policy, the diversion of critical forces to rear areas seriously impacted the

Union's inherent superiority in raw manpower. As a result, the disparity in strength on the battlefield between the Army of the Potomac and the Army of Northern Virginia did not play as significant a role as it should have, permitting General Robert E. Lee to pummel his opponents for the first three years of the war.

It has been claimed by some that the activities of partisan ranger bands in northern and western Virginia, especially those of John S. Mosby, may have prevented a Union victory in the summer or fall of 1864. Hence, it could be stated that the calculated employment of Ranger units imbued with boldness and audacity can create favorable conditions for the decisive battle elsewhere—and that is what counts in war.

Lt. Col John Mosby

There were probably but few men in the South who could have commanded successfully a separate detachment in the rear of an opposing army, and so near the border of hostilities, as long as he did without losing his entire command.
Lieutenant General Ulysses S. Grant, of Mosby in his book Memoirs

On March 9, 1863, at Fairfax Court House, General Edwin H. Stoughton, the youngest Union General at the time, was sleeping in his quarters when Confederate Raider Mosby captured him and 32 Union men. (Library of Congress)

Brigadier General Edwin H. Stoughton

SEVEN

TRAINING

May a dying soldier's last words on the field of battle never be "If only I'd been better trained."

DATE: *30 January 1944*

WAR/CONFLICT: *World War II – European Theater of Operations*

LOCATION: *Cisterna, Italy, vicinity of Anzio Beachhead*

MISSION: *To seize and hold the road junction at Cisterna until relieved.*

BACKGROUND

THE SECOND WORLD WAR saw the dawning of the modern Ranger organization (see Introduction). In the spring of 1942, U.S. Army Chief of Staff, General George C. Marshall, sent Colonel Lucian K. Truscott, Jr., to London to arrange American participation in British Commando raids against German occupied Europe. On 26 May 1942, the newly promoted Brigadier General Truscott recommended to the Chief of Staff that an American unit be organized along similar lines as the British Commandos.

During one of his trips to Washington, D.C., Truscott had discussed with Major General Dwight D. Eisenhower, chief of Operations Division War Department General Staff, the possibility of activating a U.S. commando-type unit. Eisenhower's recommendation was to use a name that was not closely associated with British Special Forces.

After some thought, Truscott decided, in honor of Rogers' Rangers, to designate the unit officially as the 1st Ranger Battalion.

The legendary William Orlando Darby was selected to command the 1st Ranger Battalion which was formally activated on 19 June 1942. The Battalion's official entry into the war occurred on 8 November 1942, as part of Operation Torch—the invasion of North Africa—though fifty U.S. Army Rangers had participated 19 August 1942 with a British and Canadian Commando amphibious assault on the English Channel port of Dieppe, France, located on the upper Norman coast approximately 105 miles northwest of Paris.

The next few months saw a dramatic change to the Ranger Battalion's organization. The most significant change occurred on 19 April 1943 when Marshall authorized the activation of the 3rd and 4th Ranger Battalions, manned by an additional fifty-two officers and one thousand enlisted soldiers. Now three battalions in size, Darby's organization would be designated the 6615th "Ranger Force (Provisional)."

On 10 July 1943, after only six weeks of training, this Ranger Force spearheaded Patton's Seventh Army amphibious landings for Operation Husky, the invasion of Sicily. Designated as "Force X," the 1st and 4th Ranger Battalions successfully secured the defended beachhead at Gela for Major General Omar Bradley's II Corps.

After Sicily was secured, the campaign continued up the Italian peninsula with a successful landing at Salerno. Any illusions of a rapid advance were soon dispelled however by the swift German reinforcement of Italy under the command of General Field Marshal Alfred Kesselring, one of Germany's most competent field commanders. With a keen tactical sense of defensible terrain, Kesselring established a series of fortified lines anchored on both coasts and integrated with the numerous mountain ridges, rivers, and towns. Slowly trading space for time, Kesselring inflicted horrendous casualties on the advancing allies, who not only bled themselves white taking one mountain crag after another, but also lived a miserable existence in the mud, snow, and rain.

For the allies, Italy was World War I with a double dose of misery. To break this excruciating stalemate, the Allies planned a turning movement through an amphibious landing at Anzio that would threaten the rear of the German Winter Line and force their displacement northward and perhaps rout the German army in Italy.

Withdrawn to rest, refit, and prepare for this invasion, Ranger Force rehearsed for the operation on 17 January 1944 as part of a 3rd Infantry Division exercise. While superiors were favorably impressed with the Rangers' spirit, élan, and enthusiasm, they were significantly concerned with noted violations of doctrine and combat techniques—weaknesses compounded by their limited pre-invasion training. Companies made excessive noise during night movements, moved while flares were fired thus giving away their location, and failed to reconnoiter likely ambush sites or take local security precautions. Additionally, the high casualty rate on Sicily had taken its toll of veterans, thus leading to an even more serious deterioration of the Rangers' fighting skills. The rehearsal and preparation did not allow for adequate time to properly train Ranger replacements to a suitable standard or correct noted training deficiencies. The eventual price paid by these Rangers for these shortcomings would prove to be cataclysmic.

Just four days after the rehearsal exercise and with little time to correct tactical deficiencies, the allies invaded Anzio before dawn on 22 January 1944. Once again leading the way, Ranger Force—reinforced with the 509th Parachute Infantry Battalion and H Company of the 36th Combat Engineer Regiment—seized the port facilities, reduced enemy defensive positions, and secured the beachhead for follow-on forces. In that the local defenses were composed of two undermanned German coast-watching battalions that had been deployed to Anzio from the Winter Line for some R&R, the Rangers encountered little resistance. By midnight of D-Day, Major General John Lucas' VI Corps had landed 36,000 men and 3,200 vehicles.

Lucas failed, however, to exploit his advantage. Cautiously moving inland only eleven miles during the next five days, VI Corps' hesitant advance provided the Germans the time and the opportunity to concentrate their forces. On 28 January, General Mark Clark urged Lucas to act more aggressively. On 29 January, Lucas finally, but belatedly, acted with a sense of urgency and issued a field order outlining a major attack.

THE LEGACY

Having anticipated the corps' order, Truscott's 3rd Infantry Division had issued its own order on 28 January. Darby's Rangers would cross the line of departure (LD) at 1 AM on 30 January to seize and to hold the town of Cisterna and its vital road junctions until relieved. The Ranger Infantry Battalions were replaced on the line the morning of 29 January, and the battalion commanders met with Darby at 6 PM later that evening to discuss the plan. The intelligence annex to the division order noted that "[i]t does not now seem probable that the enemy will soon deliver a major counterattack involving units of division size; on the other hand, the enemy will probably resort to delaying action coupled with small-scale counterattacks in an effort to grind us to a standstill."

While the 3rd Infantry Division Intelligence section (G2) was optimistic and believed that the Rangers would encounter sparse opposition, Darby's headquarters believed this optimism was misplaced. Soldiers of the Hermann Goering Panzer Division had been taken prisoner in the Cisterna area. Doctrinally, this division's presence indicated significant reinforcements and preparations for a counterattack. Consequently, the Rangers believed that enemy resistance at Cisterna would be stiff.

With Darby directing the attack from the Regimental headquarters near the LD, the 1st Ranger Infantry Battalion advanced as planned. Fifteen minutes after the 1st Battalion cleared the LD, the 3rd Ranger Infantry Battalion followed in support of the 1st. The terrain between the LD and Cisterna was flat farmland with little cover or means of concealment. Moving along a previously reconnoitered route, the 1st and 3rd Battalions used a drainage ditch to avoid enemy detection.

Preceded by an eight-man mine-sweeping party, the 4th Ranger Infantry Battalion crossed the LD at 2 AM and advanced on the Conca-Isola Bella-Cisterna road toward Cisterna. Sweeping for mines and securing the route, the battalion was followed by the Ranger Cannon Company, a platoon of the 601st Tank Destroyer Battalion and the 83rd Chemical Battalion—an assigned unit that provided the Ranger's heavy mortar support.

At 2:48 AM, four radio operators, which were to have accompanied the 3rd Battalion, reported themselves lost to the regimental headquarters. Soon thereafter, the 3rd Battalion reported that it had lost contact with the 1st Battalion halfway to the objective. Compounding that problem, there was a break in contact within the 1st Ranger Infantry Battalion itself. Approximately a half-mile ahead of the 3rd Battalion, the 1st had split in two, with three companies remaining in place and three continuing to advance. The inexperience of the replacement Rangers was beginning to show.

Quickly assessing the situation and taking command of the 1st Battalion's rear element, Captain Charles Shunstrom sent a runner back to find the 3rd Battalion. Having located the lost 3rd, the runner returned to report that a German tank round had killed its newly assigned battalion commander, Major Alvah Miller.

Despite the brief engagement, no general alarm appeared to have been raised about the presence of the Rangers. Cautiously moving toward the objective, the 1st and 3rd Battalions sought to avoid detection. Though they passed close by German positions, and enemy patrols crossed in front and on both sides, the Ranger presence apparently remained undetected as daylight approached.

Continuing to crawl through empty trenches, the lead Ranger element reached a flat, open field, which was roughly triangular in shape and 1000 yards long leading to the southern edge of Cisterna. The Rangers began running toward Cisterna in an attempt to reach the town's outskirts prior to daybreak. Approximately 600 yards from the town, they passed through a German encampment and killed a large number of the surprised enemy with bayonets and knives.

The Rangers began to cross the remaining four hundred yards to the town's edge only to be met by overwhelming fire from the town. Having detected the Rangers moving northward approximately a mile south of the triangular field, the Germans had prepared their surprise welcome. Stopped in their tracks, the Rangers returned fire as best they could from a position astride a road leading into Cisterna.

Following closely behind the three lead companies of the 1st Battalion were the 3rd Battalion and the remaining three companies of the 1st. Suppressed temporarily by two enemy tanks three hundred yards short of the embattled lead elements, the follow-on elements destroyed the tanks with bazookas. Moving forward, Shunstrom was able to reestablish contact with the 1st Battalion's commander, Major Jack Dobson, another commander new to the Rangers.

Meanwhile, on the Conca-Isola Bella-Cisterna road, the 4th Ranger Infantry Battalion attempted to reinforce its sister battalions but met heavy resistance. Stopped short of Isola Bella by heavy tank, assault gun (heavily armored mobile artillery called 'Sturmgeschutz'), automatic weapon, and small arms fire, the 4th continued its efforts to reduce the roadblock. Unable to communicate with the 1st or 3rd Battalions, Darby realized that it was critical to the survival of the lead battalions that the 4th get through. Unfortunately, Darby, the 4th Ranger Infantry Battalion, and supporting 3rd Infantry Division units were fought to a standstill. Finding itself surrounded by that afternoon, the 4th Ranger Infantry Battalion would not be relieved until the next day. The lead Ranger Force was on its own.

Outnumbered, outgunned, and now commanded by captains, the 1st and 3rd Ranger Infantry Battalions continued the fight throughout the morning of January 31st. Three German tanks appeared behind the 1st and 3rd Battalions' position and were quickly destroyed by bazookas. But the realization had sunk in. The Ranger Force was surrounded. Confined to a perimeter three hundred yards in diameter, the Rangers were deluged by a storm of automatic and small arms fire. Ranger attempts to escape the encirclement and German attempts to overrun their position were met with unrestrained ferocity on both sides.

Two hours later and still continuing to hold on, the Rangers found themselves running low on ammunition. Three companies held in reserve within the battalion's perimeter allocated half of their ammunition to the companies on the perimeter.

Growing in strength as the Rangers were being bled dry, the Germans threw the newly arrived 2nd Parachute Lehr (Training) Battalion into the fray. Armor, antiaircraft batteries, and artillery in direct fire mode rained destruction down on the entrapped Rangers.

Shortly after noon, approximately the same time that the 4th Ranger Infantry Battalion realized that all hope of assisting its sister battalions had vanished, a dozen captured Rangers were marched forward toward the center of the Rangers' position by German paratroopers with armored personnel carrier support in an attempt to gain the Rangers' surrender. Ranger marksmen shot two of the escorting German paratroopers. The Germans retaliated by bayoneting two of the Ranger prisoners. The group continued to move forward. Two shots again rang out, two Germans fell, and two Rangers were bayoneted. Rangers—mostly those new to combat or the unit—began to surrender and join the advancing German group.

With the cluster of Ranger prisoners now numbering eighty, the advance continued toward the center of the American position with the German guards shouting they would shoot the prisoners if the remaining Rangers did not surrender. The surrounded Americans fired on the group for a third time, accidentally hitting some of the prisoners along with a few Germans. The gradual surrender of individual Rangers continued despite the attempts of more resolute Rangers to stop those offering to surrender by threatening to shoot them.

Recognizing that the end was near, the surviving Rangers began disassembling their weapons, burying or scattering key parts, and destroying communications equipment. The last man to speak with Colonel Darby from Cisterna was the sergeants major of the 1st Battalion, Robert Ehalt. He informed Darby that the 1st Battalion commander was wounded, the 3rd Battalion executive officer dead, and German tanks were closing in. Concluding his conversation with "So long Colonel, maybe when it's all over I'll see you again," Ehalt destroyed the radio and continued to fight for a while longer with the few Rangers left before finally surrendering.

Broken into smaller and smaller units of resistance, some Rangers continued to fight it out only to be captured or killed man by man. By the end of the day, the 1st and 3rd Ranger Infantry Battalions ceased to exist, having been completely annihilated. Of the 767 Rangers who reached Cisterna, only six made it back to American lines, the remainder were either killed or captured.

OBSERVATION

The senior German commander in Italy, General Field Marshal Albert Kesselring, had assessed the situation correctly. Recognizing Lucas' cautious nature, Kesselring had concentrated considerable strength in the Cisterna area of operation in preparation for a German counterattack on 2 February. Even worse for the Rangers, he also anticipated the American attack on Cisterna and took measures to defeat it by reinforcing its garrison with elements of the Hermann Goering Panzer Division on the evening of 29 January.

There were two critical shortcomings that led to Ranger Force's destruction. The first, and most critical, was the failure of intelligence to determine accurately the devastating strength of the enemy the Rangers were facing. The second major shortcoming was the lack of training provided the Rangers before their commitment to combat. This lack of training impacted at both the individual and group levels. At the individual level, the newly assigned Rangers lacked those prerequisite skills that made them unique and confident light fighters. At the group level, the lack of individual skills and time to train adequately hindered the Ranger Force's ability to move smoothly and to fight adroitly as a cohesive team.

In the end, neither of these shortcomings could be blamed on the Rangers, themselves, for the solutions to each were out of their hands and, thus, beyond their ability to influence. When it came to the gathering and interpretation of intelligence, that responsibility was not theirs. As for training to standard, the Ranger standard of training was the necessary requirement to execute successfully such a daring mission. Unfortunately for the soldiers of Ranger Force, they had not been provided the time necessary to train to those Ranger standards.

Despite their destruction, however, the Ranger Force extracted a heavy price for their annihilation, inflicting over 5,500 German casualties and seriously disrupting the planned German counterattack on the beachhead, thus delaying its execution by two critical days. This additional forty-eight hours proved to be the difference for the American forces because the German counterattack failed by only the narrowest of margins as American artillery batteries were forced into the direct fire mode over open gun sights—basically looking down the barrel to target the enemy. Even clerks and cooks were pressed into service as infantrymen to repel the German counterattack.

Had the Ranger attack on Cisterna been any less forceful and unwavering, the U.S. VI Corps could have conceivably found itself conducting another Dunkirk. The Rangers' great sacrifice had thwarted Hitler's "Push the Allies into the sea!" directive. If the Germans were duly impressed by the Ranger's tenacity at Anzio, they were about to be astounded by their ferocity at their next meeting—Normandy.

LTC Darby at rest after field chow. North Africa. (12 Dec. 1942)

BRIGADIER GENERAL WILLIAM O. DARBY

(Posthumous)

On more than one occasion, Darby was offered command of a regiment and its subsequent promotion to colonel, only to turn it down.

Lieutenant General Omar Bradley, commander II Corps, observed one such occasion when Lieutenant General George S. Patton, Jr., commander 7th Army, offered the Ranger lieutenant colonel command of a regiment in his 45th Division:

"You mean I get a choice, General? I'm not used to choices in the Army."

"Take the regiment and I'll make you a colonel in the morning, but I won't force your hand. There are a thousand colonels in the army who'd give their eyeteeth for this chance."

"Well, thanks anyhow, General, but I think maybe I'd better stick with my boys."

Col. Darby awarded the Distinguished Service Medal (British) for extraordinary heroism above and beyond the call of duty. San Marco, Italy. (7 Apr. 1944)

1st Ranger Battalion loads into British Landing Craft Assaults (LCA) for beach landing maneuvers near Naples, Italy, for amphibious warfare training prior to the Anzio landings. (14 Jan. 1944)

Soldiers of the 3rd Ranger Battalion loading aboard Landing Craft Infantry (LCI) at Baia, Italy, prior to departure for the Anzio landings. (16 Jan. 1944)

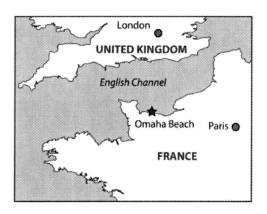

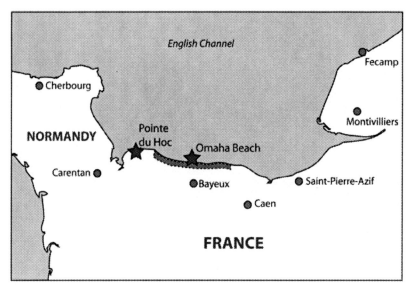

EIGHT

FEARLESSNESS

The man who conquers in war is the man who is least afraid of death.

— Alexei Kuropatkin
Russian General and Minister of War, 1848–1925

DATE: *6–9 June 1944*

WAR/CONFLICT: *World War II – European Theater of Operations*

LOCATION: *Pointe-du-Hoc and Omaha Beach, Normandy, France*

MISSION: *On D-Day, scale the cliffs of Pointe-du-Hoc to capture or destroy emplaced enemy artillery and spearhead the amphibious assault on Omaha Beach.*

BACKGROUND

THE TWO OTHER EUROPEAN Theater Ranger Infantry Battalions, the 2nd and 5th Ranger Infantry Battalions, were officially designated the Provisional Ranger Group on 6 May 1944. Under the command of Lieutenant Colonel James Earl Rudder, these two battalions—referred to by some as "suicide squads"—had their baptism of fire on the beaches of Normandy on D-Day, 6 June 1944, as part of Operation Neptune, the amphibious phase of Operation Overlord. Operation Overlord was the official title for the D-Day invasion of Normandy which encompassed both an airborne assault that commenced the night of 5 June and the seaborne assault the morning of 6 June.

Hitler's Atlantic Wall of Fortress Europe (Atlantikwall, Festung

Europa) stretched a total of 2,400 miles along the French coast of the English Channel and beyond. Field Marshal Erwin "The Desert Fox" Rommel, former commander of the Afrika Korps, commanded Army Group B which encompassed the Normandy sector, in the central-northwest area of the French coastline.

The 2nd Ranger Infantry Battalion's mission was the toughest, riskiest, and most dangerous task of any of the D-Day assaulting units: three of its companies, D, E, and F, each composed of approximately 150 men, were to land about four miles west of the right flank of Omaha Beach on a narrow strip of beach at the base of Pointe-du-Hoc, scale the 120-foot sheer cliffs under heavy enemy fire, penetrate and secure the open area, and destroy six 155-mm cannons located in reinforced concrete casemates. Fortunately, the minefield and wire obstacles were located on the landward side as protection against assaults from the rear, for the Germans eliminated a direct assault from the sea as unlikely. Lieutenant General Omar Bradley remarked of Lieutenant Colonel Rudder, "Never has any commander been given a more desperate mission."

The remainder of the Ranger force would wait offshore under the command of newly promoted Lieutenant Colonel Max Schneider. If the assault force under Rudder took possession of Pointe-du-Hoc, the code phrase "Praise the Lord" would be transmitted and the force under Schneider would move by boat to the Pointe to reinforce the position. If the message was not received by 7 AM, Schneider's force would move to the Dog Green and Dog White sectors of Omaha Beach (the U.S. 29th Infantry Division sector), move through the Vierville Draw, and proceed overland to assist their fellow Rangers at the Pointe.

Pointe-du-Hoc had already been under intense Allied air and sea bombardment for weeks. It was key terrain. From its location, the heavily dug-in but wheel-carriaged 155-mm cannon had a range estimated to be 15 miles that allowed the enemy to place fires along the full lengths of both Omaha Beach, fourteen miles distant, and Utah Beach, ten miles distant. This would not only wreak havoc on Allied ground troops as they disembarked from their assault boats but such range would provide the defending Germans an opportunity to engage larger troop and equipment transports farther out to sea.

What was believed at the time to be the coup-de-grace was delivered by the battleship *USS Texas* when it delivered salvo after salvo of 14-inch rounds on the target on 6 June shortly after a 6 AM aerial bombardment by eighteen medium bombers of the U.S. Ninth Air Force. The cumulative total of explosives delivered against Pointe-du-Hoc was calculated to be more than ten thousand tons of explosives—the equivalent of half the tonnage of "Little Boy," the nuclear bomb that would be dropped by the Enola Gay B-29 Superfortress bomber on Hiroshima, Japan, 6 August 1945.

THE LEGACY

H-Hour—the designated time of attack—was set for 6:30 AM, just past extreme low tide, on the 6th of June. General of the Army Dwight D. Eisenhower, Supreme Allied Commander, had made a fateful decision to launch the invasion despite questionable weather. High winds and rough seas were the best he could hope for during what was believed to be a limited window of opportunity between periods of gale force conditions.

From the start, the operation did not go as planned for the 2nd Ranger Infantry Battalion. Loaded aboard ten Landing Craft Assault (LCA) vessels crewed by British seamen, the force struggled against the high seas. Changes to the LCA's design made the waterborne assault even more difficult. Light armor had been added to the gunwales and sides of the boats and, while this increased the Rangers' survivability against enemy fire, it also increased the weight of the boats thus increasing the boat's depth below the water line.

Riding low in the water at slower than designed speeds, the boats began to take on water soon after launch as water washed over the sides. The force began to take losses even before the coastline came into view. The boat carrying the D Company commander was swamped, dumping him and twenty of his soldiers into the cold, numbing channel waters where they treaded water for a couple of hours prior to being rescued. A little farther on, one of the supply boats sank and a second was forced to throw half of its cargo over-board in order to remain afloat.

At 6:30 AM, the *Texas* ceased fire on Pointe-du-Hoc at the moment the Rangers were to be touching down at the cliff's base. Unfortunately at 6:30 AM, Rudder in the lead assault craft, realized that his boat's navigator—"coxswain"—was heading the group toward Pointe-de-la-Percee, 2.5 miles off target on the Omaha Beach side of du-Hoc. Fighting a strong current and running a gauntlet of German fire from dug-in positions, the tiny invasion force slowly paralleled the beach at the cost of one of their four DUKWs—a 2.5 ton cargo carrier that could make five-knot speeds in moderate seas and 50 mph overland—sunk by a 20-mm shell.

More critical than the loss of the DUKW was the impact on their time schedule, for they were thirty-five minutes late. The plan called for the Rangers to hit the base of the objective just moments after the *Texas* lifted her heavy suppressive fires that were planned to drive the Germans into the protection of underground bunkers. It was hoped that the Rangers could scale the cliff before the defenders could reoccupy their defensive positions.

The original plan called for the three D Company LCAs to land on the west side but, because of the navigation error and the loss of one of the D Company boats, Rudder elected to have all nine LCAs land side by side on the east side of the Pointe. The enemy fire was intense as the Rangers approached. Machine gun and cannon fire from the front and flanks ripped into them as Germans reoccupied their positions along the edge of the cliff and fired down on the advancing Americans. Two destroyers, the *USS Satterlee* and *HMS Talybont*, repositioned offshore after observing what was happening to the attacking Rangers. By providing 5-inch gun, 40-mm, and 20-mm fire support, these two Navy warships were able to drive some of the defenders from the cliff's edge.

The beach was only 30 feet wide at the point of attack and was literally losing ground with each passing minute as the tide rose. Composed mainly of gravel, not sand, the beach would be completely submerged by high tide. Additionally, the heavy artillery pounding against the cliff caused huge chunks of soil to crash down the cliff.

This cliff debris, mainly clay, only served to make the shale rock slipperier. There was one silver lining, though. The soil accumulation at the base was over twenty feet high in some areas thus giving the climbing attackers a height advantage.

The Rangers had a number of inventive devices to assist their scramble to the top. Unfortunately, much of that special mountain climbing equipment did not make it to the shore having been lost or jettisoned at sea. Of the equipment that reached the beachhead, adverse weather conditions seriously hindered its utilization. One device that successfully landed was a seventy-five foot assault ladder. These donated extension ladders from the London Fire Department were mounted on the DUKWs. The three surviving DUKWs were unable to get traction on the clay-covered shale and were thus unable to get close enough to the cliffs to extend their ladders against the rock face. Despite the rolling effect of the swells, however, one DUKW was able to extend its ladder while afloat. Heroically, Sergeant William Stivison climbed to the top of the ladder, from where he began to engage the enemy on the cliff with his machine gun. Swaying widely back and forth in a forty-five degree arc to both sides of vertical—the extension basically serving as a huge moment arm that threatened to sink the DUKW—the ladder had to eventually be brought back down.

Another plan called for steel grapnels—multi-pronged hooks—to be fired from the LCAs. Fired just prior to each LCA touching down, three rocket-propelled grapnels trailing either a rope ladder, plain three-quarter-inch rope, or toggle ropes exploded from the decks of the boats and arced their way toward the cliff. Attached to each grapnel was a burning fuse lit just before launching to make the Germans believe that the grapnels were some type of exploding device that should be avoided. The Rangers watched with great apprehension as the water-soaked lines followed behind the rockets. Though a majority of grapnels failed to reach the cliff, at least one from each boat did.

Unfortunately for the Rangers, a second problem was encountered in regard to the bombardment of Pointe-du-Hoc. Bombs and shells that had fallen short had created craters underwater. Unseen

because they were submerged by the rising tide, these craters swal-
lowed many a Ranger as he struggled off the boat ramps. One Ranger, Lieutenant Kerchner, was resolved to be the first Ranger to hit the
beach from his boat. Jumping into what he believed to be three feet
of water with a shout of "OK, let's go," the lieutenant found himself
literally in over his head as he leaped into a large shell hole. Noting
their lieutenant struggling to remain afloat, the men following exited
to the sides, and advanced to shore hardly getting their feet wet.

Each Ranger company had an authorized strength of three officers and seventy-four enlisted, only about half that of a traditional infantry company. One of the first assault craft to make it to the Pointe's
beach was Colonel Rudder's. Disembarking from their beached boats,
Rudder's men assaulted across the gradually narrowing strand of
land to the base of the cliff. Taking heavy machine gun fire from a
position located on their left flank, the Rangers quickly lost fifteen
men, dead or wounded. Reaching the ropes, the surviving Rangers
began their arduous climb up the suspended rigging.

The Rangers made the climb with various degrees of success and
difficulty. For some of them, the climb proved to be much less difficult than their practice climbs back in England. For others, wet and
clay-covered ropes required a number of efforts to negotiate successfully.

Not anticipating an attack on their position from the sea, the German defenses facing the Rangers were the weakest. As the Rangers
climbed, the Germans tossed grenades over the cliff and cut three or
four of the ropes. Ranger Browning Automatic Rifle (BAR) fire from
below and continued supporting fires from the USS Satterlee drove
many of the defenders from the cliff's edge. Within five minutes of
starting the climb, the first Rangers were secure on top to be followed
in another ten minutes by the bulk of the fighting force.

At approximately 7:30 AM, the message "Praise the Lord" was
broadcast from Pointe-du-Hoc, indicating that the Rangers had successfully scaled the cliffs. With less than two hundred Rangers left to
continue the attack on top of the cliff, Colonel Rudder's signal to his
floating reserve, A and B Companies, 2nd Ranger Infantry Battalion,
and the 5th Ranger Infantry Battalion to reinforce at Pointe-du-Hoc

was transmitted fifteen minutes too late. The reserves had moved on to assault Omaha at 7:15 AM in accordance with their established contingency plans.

The top of the cliff was like nothing the Rangers had ever seen before. Having studied in great detail maps, sketches, and aerial photos of the objective, they were amazed to find "just one large shell crater after the other." Large pieces of concrete, blown off the reinforced casements (fortifications), were strewn about the area.

Each platoon had a specific mission and despite the losses, despite the heavy concentration of defensive fire, and despite the intermingling of units, each moved out as rehearsed and without being ordered to do so. For once, the shell holes and craters turned out to be an asset to the Rangers, providing immediate cover and concealment. Ignoring the machine gun and 20-mm cannon fire directed at them, the Rangers moved through the shell craters and German trenches toward the six heavy gun casemates. Closing on their mission objective, they were stunned to find that the "guns" were, in reality, telephone poles. Fresh tracks running inland indicated that the six weapons had been recently removed.

Undaunted and unfazed, the Rangers continued moving inland in small groups toward their second objective, the hard surface road that linked the towns of Vierville and Grandcamp. Intent upon establishing a roadblock across the road to prevent German reinforcements from moving to Omaha Beach, the Rangers encountered heavy resistance near the outer perimeter of Pointe-du-Hoc. The Pointe was a fort and the landside perimeter was heavily defended with machine gun emplacements, barbed wire entanglements, and antipersonnel minefields. By the time D Company was able to establish a blocking position across the road, only twenty Rangers remained combat-effective out of the seventy who had loaded on the assault boats.

Though the Rangers had swept across the objective and established a roadblock along the Vierville and Grandcamp road, the situation was far from stabilized. A battalion aid station was established in a two-room concrete fortification and the number of Ranger wounded quickly overwhelmed its capacity. Movement above ground was still nearly impossible so the Rangers stuck to the trenches. Germans, moving by underground passages, continued to snipe then quickly disappear. Individuals and small groups carried on the fight, with both sides taking prisoners.

Yet to be silenced was the machine gun strong point on the eastern edge of the German position that had caused so many casualties during the initial attack across the small beach below. Ordered by Rudder to eliminate it, Lieutenant Elmer "Dutch" Vermeer moved forward with a small patrol. Moving through shell holes, Vermeer encountered a patrol from Company F determined to accomplish the same mission. Unexpectedly, as both patrols continued to move forward, the craters disappeared, leaving a flat, open space of nearly three hundred yards to assault across.

Realizing that their losses would be heavy, Vermeer prepared to move forward in a frontal attack just as Rudder ordered the patrols to hold their position. Lieutenant Eikner, the colonel's communications officer, had brought with him an old World War I (WWI) signal lamp with shutters. Having lost their naval shore-fire-control party earlier when it was obliterated by a friendly artillery round that fell short on it and having lost all other means of radio communication during the initial assault on the cliff, Eikner decided to use the signal lamp to contact the naval destroyer USS *Satterlee* by Morse code and have the ship train its five-inch guns on the pesky machine gun position. Locating enough dry-cell batteries to get the device working, he established two-way communications with the destroyer. Shortly thereafter, the *Satterlee* blew the machine gun position away with a direct hit.

Despite the naval gunfire support, the Rangers still found themselves in a very precarious position; they were cut off from the sea as intense enemy fires still prevented any additional support ships from landing. They were still cut off by land, as Allied forces had not yet been able to force the Vierville Draw on Omaha Beach. They had no radio communications, other than the old lantern, for all of their equipment had been either lost, destroyed, or damaged. And now, despite the fact that they had lost over half of their men, the Rangers were still virtually isolated.

Furthermore, to add insult to misery, the British cruiser *HMS*

Glasgow had fired a short marker round—filled with dye rather than with explosives but still packing a punch—that struck in the vicinity of Rudder's command post (CP), knocking him senseless, killing one, and wounding another. Recovering quickly, Rudder angrily went hunting for snipers only to be shot in the leg. Fortunately for the Rangers and despite the pain, the force commander continued to direct his unit's defense. Vermeer concisely stated Rudder's contribution to the operation. "He was the strength of the whole operation."

At the roadblock, the fighting continued with much of it at close quarters. Companies D and E had the roadblock perimeter established by 8:15 AM with a force of thirty-five Rangers. Twelve Rangers from Company F joined them fifteen minutes later. Patrols were immediately sent out to reconnoiter the area.

Following heavy tracks of the missing artillery pieces for nearly 250 yards down a dirt road, Sergeants Leonard Lomell and Jack Kuhn abruptly stopped. They had located the five missing cannons (though some reports indicate there were six). Well camouflaged and prepared to fire on Utah Beach, the weapons were stocked with piles of ready ammunition but no gun crews. Having apparently moved away from the guns during the heavy pre-invasion bombardment for fear of having their own ammunition cache hit, the one hundred Germans that constituted the crews were only now beginning to form across an open field approximately a hundred yards away. They appeared to be in no hurry, for they had to wait until the observation tower on Pointe-du-Hoc was recaptured in order to place accurate and adjusted fires on the beachheads.

Never hesitating, Lomell secured some thermite grenades—hand held devices designed to destroy equipment with extreme heat—Kuhn was carrying and moved to the guns. The grenades were placed in the traversing and recoil mechanisms of two of the guns. Then, prior to withdrawing, he destroyed the sighting mechanism of a third. Racing back to the roadblock, Lomell and Kuhn collected more thermite grenades which they ignited in the traversing and recoil mechanisms of the remaining guns.

At the same time, a patrol led by Sergeant Frank Rupinski stumbled across a huge stockpile of ammunition south of the gun battery's location. Using explosives, Rupinski destroyed the ammunition dump. By 9 AM, word was quickly sent back to Rudder by messenger that both the guns and ammunition were destroyed. Mission complete: the first United States forces to accomplish their mission on D-Day.

For the next forty-eight hours, the Rangers were relatively on their own and still isolated. Armed with nothing heavier than BARs and 60-mm mortars, the force did receive some reinforcements in the form of a platoon of Rangers from Omaha Beach. Led by Lieutenant Charles Parker, the twenty-three soldiers arrived at 9 PM that evening with an additional twenty POWs in tow. During the afternoon of the 7th, a landing craft was able to remove the wounded and prisoners of war as well as bring a reinforcement of twenty Rangers from the 5th Battalion from Omaha Beach.

The siege of Pointe-du-Hoc continued as the Germans attempted to retake the fortified area with a series of counterattacks throughout the 6th of June and into the next day. But, despite their desperate situation, the Rangers held. By the end of the battle on 7 June, having fought nonstop for two days without relief, only 25 percent of the original force—fifty Rangers—was still capable of fighting.

Some historians claim that the mission was all for naught, that the unit scaled the heights and secured the Pointe at such a terrible cost for nothing. That claim is not substantiated however. While the guns were not located and destroyed on the cliff, they were located in the vicinity, ready to fire on the struggling and congested beachheads. Their added explosive weight to the fray below could have been significant. At the worst, it would have proved devastating to Omaha Beach, just in terms of morale alone. At the best, it spared many more Allied troops and much equipment, at a relatively smaller cost to the attacking Rangers. Regardless of their eventual location, the guns still could not have been located and destroyed prior to their opening fire without the Rangers scaling the cliffs.

Ten years later, during the reunion at Normandy, Colonel Rudder revisited the site with his son. At the base of the Pointe, he looked up the towering cliff and asked, "Will you tell me how we did this?"

As Companies D, E, and F prepared to take Pointe-du-Hoc, the remainder of the 2nd Ranger Infantry Battalion, A, B, and C Companies, maintained their position in the channel. Companies A and B waited with the 5th Ranger Infantry Battalion for word from Rudder's Rangers at Pointe-du-Hoc—word to either reinforce the 2nd Battalion on Pointe-du-Hoc or move on to Omaha Beach. C Company was prepared to land at H+03 minutes—three minutes after the start of the invasion—in sector Dog Green of Omaha Beach on the far right flank of the assaulting units, only two minutes after the initial landing of Company A, 116th Infantry Regiment (A/116th).

The 116th Infantry Regiment was a Virginia National Guard unit and the success of C Company's mission depended significantly on A/116th Infantry crossing the beach and securing the Vierville Draw at the start of the invasion. Unfortunately for the C Company Rangers, A/116th Infantry company was virtually annihilated prior to crossing the beach.

Approaching Dog Green in six LCA assault boats at H+1—one hour after the start of the invasion, A Company was initially engaged by enemy shore batteries 5,000 yards from the beach. No casualties were sustained until LCA 5 received a direct hit at 1,000 yards and six infantrymen drowned. LCA 3 took a hit at one hundred yards, killing two more soldiers outright and drowning another twelve as the boat sank. At 6:36 AM, the remaining boats dropped their ramps. Water depth ran from chest height to over one's head. Machine gun fire swept the entire length of the amphibious line. Even lightly wounded soldiers drowned as their heavy soldier's load drove them under the waves. Survivors crawled along the water's edge, attempting to obtain some sort of concealment from the churning foam. Fifteen minutes after the first ramp had dropped, A Company had ceased to exist as an effective fighting force. Within thirty minutes, nearly two thirds of the unit, including all officers and most NCOs, was either killed or wounded.

Immediately behind A/116th in two LCAs and unable to see any of the carnage on Omaha, the Rangers of Company C felt rather confident that they would meet little resistance at the water's edge—considering all of the air and sea bombardment that had transpired earlier. This confidence soon transformed itself into consternation as they realized they were off course and about to be the first to assault the section of beach they were about to land on. Moments later at 6:45 AM, the Rangers hit the shore at the far western edge of Omaha Beach, out of position just west of the Vierville Draw—a narrow valley, a natural passage that made its way towards the beach.

More than two thousand yards from the nearest supporting troops, the Rangers of Company C were isolated and on their own. But, they were not alone. For, as the first LCA's ramp dropped, the German defenders opened up with a lethal concentration of rifle, machine gun, mortar, and artillery fire. Machine gun fire ripped through the thin-skinned boat and across the water, killing many of the first ones who entered the water. With little chance of survival exiting the front of the boats, many Rangers elected to jump over the sides. Loaded down with rifles, grenades, ammunition, radios, bedrolls, and personal gear, many of the Rangers from the first LCA found themselves sinking below the waves of the channel, unfortunate enough to have had their LCA hit a sand bar some distance from the shore's edge. Mortar and artillery rounds created large geysers of water as the Germans targeted the second LCA, which was hit three times by artillery fire.

Struggling through—or under—the water, individual Rangers fought through the channel surf and heavy German fires toward the water line, hoping to attain safety at the base of a bluff a few hundred yards inland. Many did not make it. Dead Rangers lay everywhere, in the destroyed LCAs, floating in the water, rolling on and off the beach as the surf ebbed and flowed. Others were strung across the beach from the water's edge to the bluff. Sergeant Walter Geldon was one of those soldiers. Lying on the beach, Geldon died the day of his third wedding anniversary. Sixty-eight C Company Rangers embarked on the two LCAs; only thirty-one fatigued men made it across the beach to the cliff. Behind them were nineteen dead and eighteen wounded. C Company's initiation to combat in the Second World War had resulted in fifty-five percent casualties and the company had yet to fire a shot!

For those Rangers who survived and made it to the base of the cliff, exhausted, their stringent, demanding, and repetitive training took effect. As grenades rolled down the cliff from overhead and impacting mortar rounds walked their way toward the base, the company commander, Captain Ralph Goranson, and his two platoon leaders, Lieutenants William Moody and Sidney Salomon, began to rally their troops. To remain on the beach meant eventual destruction. The Vierville Draw had not been secured and was heavily defended. The only viable alternative was to scale the ninety-foot cliff overhead (to the right of Vierville).

Having landed with some specialized climbing equipment, the two platoon leaders and Sergeants Julius Belcher and Richard Garrett quickly moved along the cliff's base where they located a crevice. Inserting their bayonets to establish handholds, the four Rangers used their highly honed climbing skills to scale the bluffs. From the top, they dropped toggle ropes for the remainder of the company to follow. Company C became the first Omaha Beach assault unit to reach the high ground.

The company's mission was to move west, away from the draw. But Goranson elected to send a patrol to attack the Germans, securing the draw in an attempt to provide an exit off the beach for the trapped invasion forces. Centered on a Norman stone farmhouse from which the Rangers were being engaged, the German defenses consisted of numerous pillboxes, 20-mm cannon, and a maze of communications trenches. Leading the patrol, Lieutenant Moody kicked in the farmhouse door, killed an officer, and began to move through the trenches when he was killed by a round through his forehead.

Lieutenant Salomon moved forward to continue the clearing operation. At one point, Sergeants George Morrow and Belcher located one of the machine guns that had placed such devastating firing down on their company as they attempted to disembark earlier on the beach. Incensed and infuriated, Belcher openly charged the position. Kicking in the pillbox's door, he tossed in a white phosphorus grenade and proceeded to shoot each German abandoning the position in burning agony.

Throughout the day, the battle just west of the draw raged without a decision. Unlike the Utah Beach area, airborne forces had not been inserted behind the beachhead to block reinforcements. Consequently, the Germans were able to bring fresh troops continuously into the area through their trench system, reoccupying those areas previously cleared by the Rangers. With his only reinforcements being a section of soldiers from the 116th Infantry Regiment landing a kilometer off course, Goranson was too weak to dislodge the defenders and his force was too few in number to secure the areas he did clear. To further compound their troubles, the Company C Rangers found themselves on the wrong end of some allied 20-mm and 5-inch gunfire support when U.S. Navy destroyers fired on the cliff positions without realizing that the Rangers had scaled the heights.

To many Rangers, it seemed as if the day would never end. Isolated and without radio communications, the Rangers and the few accompanying 116th soldiers continued the fight that gradually provided attacking elements the opportunity to force the Vierville Draw.

Though C Company never started, much less completed, its mission, its firefight at the Vierville Draw proved to be critical for it redirected German reinforcements and fires that would otherwise have been directed at the American force on "Bloody Omaha." Burial party reports recorded that sixty-nine German and two American dead were located in the vicinity of the stone farmhouse and its defensive trench network.

While the 5th Ranger Infantry Battalion's experience was less dramatic, it was certainly not without its own difficulties. Activated at Camp Forrest, Tennessee, on 1 September 1943, the 5th's training started 14 September. Following a plan of instruction similar to the 2nd's, the battalion trained in Florida and New Jersey prior to deploying to England early January 1944. Arriving in Liverpool on 18 January, the 5th Rangers continued to follow the same schedule as the 2nd Rangers: cliff assaults and the British Assault Training Centre.

The 5th Rangers and the two attached Ranger companies from 2nd Battalion rode the channel tides waiting for Rudder's message from Pointe-du-Hoc. Like the 2nd Battalion companies, Schneider waited until the appointed time of 7 AM for the "Praise the Lord" message. The time came and went but the Ranger continued to hold his position. Finally, at 7:15 AM, he could delay no longer. The flotilla of twenty LCAs turned to make its run at the beach.

Fifteen minutes later, the message came. But it was too late. The Ranger reserves were committed. Five minutes ahead of the 5th Battalion main body, companies A and B in five LCAs hit the beach first on the boundary between Dog Green and Dog White, just east of the Vierville Draw.

Storming Omaha Beach was the main responsibility of the American Army. Situated centrally on the Normandy invasion front, nearly double the numbers of forces were committed here than were at the other U.S. objective of Utah Beach. Omaha Beach was unlike any other in the region. Stark cliffs ranging in height from 100 to 170 feet commanded much of the water's edge. A shingle embankment about eight feet high and extending approximately fifteen feet inland lay just above the high water mark. This embankment rose, on average, at a steep rate of one vertical foot for every six feet traversed—a 16 percent slope. Though not significantly steep, this mild steepness along with the four- to six-inch-diameter "pebbles" that covered the slope's surface, conspired to create a serious obstacle to traction for both wheeled and tracked vehicles.

The beach itself was two hundred yards deep at the center and tapered off to become quite narrow at each end, enclosed by high cliffs as bookends. Along the entire rear of the designated Omaha Beach was a masonry seawall that ran in height from four to eight feet. The combination of seawall and commanding ridgeline to the immediate rear of the beach made the four primary entry/exit points to the beach—designated D-1 (Vierville Draw), D-3, E-1, and E-3—key terrain that needed to be seized if vehicular traffic was to move off the beach. Vierville Draw, in particular, was the most critical of the four since it was the only paved route and provided the shortest access to a coastal highway that ran parallel to the beach.

Recognizing the criticality of the draws, the Germans constructed twelve heavily fortified positions along the heights overlooking Omaha Beach, angled towards and concentrated on the four beach exits. The vast majority of gun positions were selected to bring nearly the entire beach under their interlocking fires. Wing walls—a form of concrete blinders constructed around the opening of a gun position—were installed to hide muzzle flashes. While these walls limited the weapon's ability to fire at targets offshore, they did not limit the weapon's ability to target landing craft closing on the shore. Ultimately, these walls proved to be an excellent investment of resources, for they significantly hindered the ability of supporting naval ships to target and destroy the casemates.

During the weeks leading up to D-Day, air and sea preparatory bombardment had been minimized along the coast prior to the invasion as a trade off to achieve surprise. The plan, though, did call for one massive sea and aerial bombardment to commence at 6 AM, just thirty minutes prior to H-Hour. Moving into position, a force composed of two battleships, four light cruisers, and twelve destroyers commenced the shore bombardment ahead of time at approximately 5:30 AM when the battleship *USS Arkansas* was spotted and engaged by a German shore battery.

Conditions and intelligence were certainly lacking. Smoke and haze prevented spotter planes from accurately locating targets and, despite intense pre-invasion analyses of aerial photographs, many of the fortified positions defending the draws were not identified as targets. The final naval prep fires—pre-planned artillery fired before the launch of most operations—were delivered by specially modified Landing Craft Tanks (LCT). As the assault craft approached in the final ten minutes prior to H-Hour, each of these LCTs launched approximately 1,000 five-inch rockets at the shore. Within that ten-minute interval, over 10,000 rockets impacted on Omaha.

Unfortunately, a military miscalculation of the first order occurred in the attempted aerial bombardment. Weather conditions made altitude bombing by heavy bombers of the U.S. Eighth Air Force exceptionally difficult that morning. Unduly concerned with dropping large amounts of ordnance on advancing assault craft, the air command elected to delay by a few seconds the release point over Omaha. The end result was a saturation bombing approximately three miles inland—not a single bomb landed on the defenses. It was a mistake that not only led to significant American casualties but, more critically, it was a mistake that almost resulted in the withdrawal of American forces from the beachhead.

Intelligence failures also played havoc with the assault forces on Omaha Beach. Reports indicated that only one reinforced battalion of the German 716th Infantry Division was defending both the Omaha and Utah Beach sectors. The defenders were believed to be trained to a minimum standard with no reserves available for counterattacks. This assessment was in total error. Despite the frequent intelligence provided by the local French resistance, an experienced German infantry division, the 352nd, had deployed to the coast early that spring and assigned a sector that was inclusive of Omaha Beach. The net result was a doubling of forces in the area, a soldier of higher standard, improved command and control, and a mobile reserve positioned only five miles from the beach. This ominous oversight was another factor that nearly led to defeat for the Americans attacking Omaha.

While Rudder's C, D, and E Companies were battling for their lives on Pointe-du-Hoc, his A and B Companies were making their way behind the first waves of assault boats to strike Omaha beach. Circumstances relating to their angle of approach, the geography, and the motion of the waves, led most of these Rangers to believe the coast was lightly defended. A Company commander, Captain Dick Merrill, even called out, "Fellows, it's an unopposed landing." How wrong he was, though, for unseen before them, the armor and infantry units from the initial waves were bogged down and scattered about at the water's edge pinned down by heavy and vicious fires. Unable to advance a single inch inland without being cut down, the initial wave of survivors were attempting to burrow their way into the sand, gravel, and shale beach in search of cover.

Hunkered down behind the walls of their assault craft, the Rangers could see none of what was in store for them until the boat ramps dropped and they began to rush forward. By then, it was too late. German defenders opened fire, unloading on the disembarking Rangers with withering fires so intense that some Rangers took nearly thirty minutes to struggle through the water to reach the beach. The casualties they incurred were almost as great as those of the 29th

Infantry Division's lead element, the 116th Infantry Regiment, which had been severely mauled and virtually eliminated within minutes of hitting the beach.

Offshore, waiting his turn to assault, the commander of the 5th Battalion, Lieutenant Colonel Schneider, observed the near destruction of the 2nd Ranger Infantry Battalion companies. Realizing that the initial landing effort had been a disaster, he elected not to sacrifice his own battalion. Ordering the remaining fifteen LCAs to move east down the coastline, he found a quieter, relatively speaking, sector of Dog Red and commenced his assault—undoubtedly saving many Ranger lives in the process.

The 5th Battalion landed at 7:45 AM on a sector of beach strewn with the wounded and dead of 116th Regiment assault teams. Leaderless, unorganized, and shell-shocked, the first wave assault teams had been under constant and heavy fire since 6:30 AM and were gripped by fear. Four hundred and fifty men of the Ranger Infantry Battalion ran across the beach toward the seawall. Of that total, four hundred and forty-four safely made the dash.

The 5th Battalion's Ranger Chaplain was Father Joe Lacy. The evening prior, he had advised his flock, "When you land on the beach and you get in there, I don't want to see anybody kneeling down and praying. If I do, I'm gonna come up and boot you in the tail. You leave the praying to me and you do the fighting." The next day, while the Rangers fought, the good father moved about the shoreline, hauling the dead, dying, and wounded from the water to protected positions where he could either provide comfort or tend to last rights.

Enemy resistance had greatly exceeded expectations. By 8:30 AM, all landings at Omaha Beach were halted and Lieutenant General Omar Bradley was seriously considering redirecting the follow-on forces to one of the other beachheads. Over 5,000 soldiers were trapped on the beach with nearly 50 percent of them lying wounded or dead on the shore. Body parts—headless torsos, arms, legs—floated in the surf. Groups of exhausted, confused, and frightened soldiers huddled wherever they could find some cover or concealment. Equipment was stacked on the waterfront, and vehicles that were not already hit and burning were immobilized, serving only as inviting targets for the ranging mortar and artillery rounds.

Though it would seem at that moment that the Allied war had been lost on Omaha Beach, there was American perseverance and some movement. Tanks of the 741st Tank Battalion, the first to arrive at Omaha, put up a stiff fight despite their inability to negotiate the rocky, pebbly beach. Caught between rising waters and German anti-armor fires against their exposed positions, the tankers continued to engage the over-watching enemy pillboxes until either taken out of action by a hit or flooded out by the incoming tide. Individuals and small groups bypassed the heavily defended beach exits and began to move directly away from the beach by moving straight up the vertical cliffs in front of them. Bravely moving onward, these courageous soldiers passed "many dead bodies, all facing forward."

Realizing that the situation was critical and that his forces must clear the beachhead, the assistant division commander of the 29th Infantry Division and first general officer to the beach, Brigadier General Norman D. Cota, set about moving his troops inland. Cota served as an inspiration to all who saw him on that bloody day. Shortly after arriving on the beach, he gathered a group of men and led it through a mortar barrage across the beach and up the bluff. Reaching the top of the bluff at a point midway between St. Laurent and Vierville and meeting some resistance, Cota organized his group into fire and maneuver teams that drove the German defenders to flight.

Progressing along a dirt road that ran parallel to the beach, the general moved through and secured the town of Vierville. On the western edge of the town, he dispatched a newly arrived twenty-three-man patrol of Rangers towards Pointe-du-Hoc. Encountering stiff resistance, the Rangers' movement was nearly stopped until Cota moved forward to assist the Ranger platoon leader Lieutenant Parker, with the deployment of his men. This group of Rangers would be the group that would link up with the 2nd Battalion at Pointe-du-Hoc around 9 PM.

Needing to get back to the beach to provide senior command and control, Cota moved to the Vierville Draw accompanied by his aide and four riflemen. Still heavily defended by German troops, the draw had just finished being pounded by the *Texas's* secondary armament of 5-inch batteries when Cota's small group arrived to find German troops quickly moving from their bunkers and reoccupying their positions. Observed and fired on by some of the defenders, Cota's group was able to capture five German prisoners who showed them a safe passage through the minefields of the draw, allowing them all to reach the beach safely.

Under constant machine gun and sniper fire from the bluffs, Cota continued to move about the beachhead, ordering, cajoling, herding, and reorganizing units, telling his soldiers, "Don't die on the beaches, die up on the bluff if you have to die, but get off the beaches or you're sure to die." Moving from group to group, he came across one of his sons' West Point classmates, Captain Raaen, commander of the 5th Ranger Infantry Battalion Headquarters, Headquarters Company (HHC). Directed to Schneider's CP, the general commented to Raaen, "You men are Rangers and I know you won't let me down."

Locating the Ranger Infantry Battalion's CP, Cota remained standing, prompting Lieutenant Colonel Schneider to stand to speak with the general. Cota asked of the men around him, "What unit is this?"

Schneider replied, "We're Rangers, sir!"

With that, Cota yelled, "Rangers, lead the way off this beach before we're all killed." Thus was born what would eventually become the official motto and mantra of the 75th Ranger Regiment: *Rangers Lead the Way!*

Schneider and his Rangers were in the final stages of preparation to break out of the beachhead when Cota arrived at their location. Moving forward, Corporal Gale Beccue of B Company and an accompanying private, shoved an M-1A1 Bangalore torpedo—a five-foot section of steel tube filled with TNT and amotal—under some barbed wire to blow a gap in the obstacle.

Lieutenant Francis W. "Bull" Dawson was tossed on top of a barrier wall by some of his men. Charging through more wire, Dawson destroyed a machine gun nest, cleared trenches of German defenders, and secured prisoners. The lieutenant's efforts inspired the rest to follow. For his actions, Dawson was awarded the Distinguished Service Cross.

Having breached the main German defenses on the beach, the Rangers then advanced up the bluff, encountering little opposition along the way. However, the push was not without its moment of modest relief. For a brief period, members of the Headquarters Company donned their chemical protective mask in belief that the heavy smoke from a brush fire might be poisonous gas. Some men nearly suffocated, having forgotten to pull the plug on the front of the mask that allowed air circulation for normal breathing. For sure, that only happened once!

Cresting the top, Schneider deployed his battalion to the left and right flanks and dispatched a unit to move four miles down the road to seize Vierville. To the front lay open fields and a maze of hedgerows from which German machine guns engaged the Rangers—another significant drawback of failed aerial bombardment earlier that morning. Breaking up into smaller assault groups, the Rangers and other surviving members of the 116th Infantry Regiment who had also made it to the top had to move forward across intermittently open ground to engage and to outflank the enemy defensive positions. The breakout from Omaha Beach had begun.

Finally, on D+3—three days after the invasion, the 5th Ranger Infantry Battalion linked up with its sister battalion at Pointe-du-Hoc. Though their losses were less than other assault elements at Omaha, the 5th Rangers suffered 23 Killed in Action (KIA), 89 Wounded in Action (WIA), and 2 Missing in Action (MIA)—25 percent casualties—during the first five days of fighting in Normandy.

OBSERVATION

Normandy was the greatest amphibious invasion and, arguably, the most challenging Ranger mission in history. History fails to give due credit to the soldiers assaulting the beaches because the enormous materiel and firepower superiority of the Allies suggests that victory was preordained. Since the aerial and naval bombardments had failed to suppress, much less destroy, the German positions, the onus of success fell on the shoulders of the men storming the shoreline. Given the German immediate advantages—and they were very apparent to the men struggling ashore—the Americans could have easily lost their nerve and surrendered for their position certainly looked hopeless. The perseverance and grim determination of small groups of soldiers fostered a fearlessness that created the first cracks in the German defense, which widened as more soldiers followed the successful penetrations.

War challenges the human dimension and often the line between giving up and perseverance comes down to a few determined men. Over the course of those eventful days, events proved that the Rangers of the 2nd and 5th Ranger Infantry Battalions were up to the challenge, though their fearlessness and incredible bravery ultimately would leave many of them buried in the fields of France. Their sacrifice, however, would establish the standard for all future American warriors to emulate.

In Flanders Fields

In Flanders fields the poppies blow
Between the crosses, row on row
That mark our place; and in the sky
The larks, still bravely singing, fly
Scarce heard amid the guns below.
We are the Dead. Short days ago
We lived, felt dawn, saw sunsets glow,
Loved and were loved, and now we lie
In Flanders fields.
Take up our quarrel with the foe
To you from failing hands we throw
The torch; be yours to hold it high.
If ye break faith with us who die
We shall not sleep, though poppies grow
In Flanders fields.

John McCrue (1872–1918)

2nd Battalion Rangers working their way up the cliff of Pointe du Hoc after successfully securing the top. (6 Jun 1944)

2nd Battalion Rangers with German prisoners captured at Pointe du Hoc. (6 Jun. 1944)

One of the five German artillery pieces discovered and destroyed by the 2nd Battalion Rangers. Surprised to not find the cannon on Pointe du Hoc, a small team of Rangers on a scout mission located them nearly a half mile away. (6 Jun. 1944)

A 2nd Battalion Ranger aid station at Pointe du Hoc. (6 Jun. 1944)

This picture is of 1st Battalion Rangers rehearsing their duties loading assault landing crafts (LCA) from an infantry landing craft (LCI) while enroute to the North African Theatre of Operations. (6 Nov. 42) 5th Battalion Rangers, along with elements of the 2nd Battalion Rangers, would have conducted similar operations off the coast of Normandy on D-Day.

This picture is of an assault landing craft (LCA) after being lowered into the water to carry Rangers to the Algerian shore in North Africa. (8 Nov. 42) 5th Battalion Rangers, along with elements of the 2nd Battalion Rangers, would have conducted similar operations off the coast of Normandy on D-Day.

5th Battalion Rangers, along with elements of the 2nd Battalion Rangers, would have had a similar view as they approached Omaha Beach on 6 June 1944

This is what members of the 5th Battalion Rangers and elements of the 2nd Battalion Rangers saw in the initial moments of the amphibious invasion when the assault craft's (LCA) ramp dropped leaving them to wade through the water and enemy fire to make for the beach.

NINE

ENDURANCE

He conquers who endures.

— Aulus Persius Flaccus
Roman satirical poet, AD 34–AD 62

DATE: *23–27 February 1945*

WAR/CONFLICT: *World War II – European Theater of Operations*

LOCATION: *Vicinity of Irsch – Zerf, Germany*

MISSION: *To infiltrate deep behind enemy lines to establish and defend a blocking position on a critical road until relieved.*

BACKGROUND

WITH THE WAR IN Europe coming to a close and Allied forces advancing rapidly, the 5th Ranger Infantry Battalion, under the command of Lieutenant Colonel Richard Sullivan, was providing long-range patrols and other conventional infantry operations including security for the 12th Army Group Headquarters in October and November 1944. During the first eighteen days of December, while attached to the 6th Cavalry Group, the battalion suffered an additional 18 KIA, 106 WIA, and 5 MIA clearing towns.

Seriously weakened by the steady loss of experienced Rangers, the 5th Ranger Infantry Battalion was gradually becoming an undermanned and depleted conventional light infantry battalion. On January 1945, the battalion was again conducting security missions, but the 5th's luck would soon change for the better with their first assignment of a true Ranger-style mission.

On 22 February 1945, elements of Major General Walton H. Walker's XX Corps crossed the Saar River in the Serrig and Taben areas. In a move designed to quicken the expansion into enemy territory by encouraging a German withdrawal, Walker gave the 5th Ranger Infantry Battalion on 23 February the task of infiltrating deep behind enemy lines in order to block the critical Irsch-Zerf road. This mission would prove to be the last major combat mission of any Ranger Infantry Battalion in the Second World War.

THE LEGACY

At noon on 23 February, the 5th Ranger Infantry Battalion was attached to the 94th Infantry Division and immediately set about planning the mission. Minus B Company, the battalion assembled in Weiten to replenish with extra ammunition, antitank mines, and rations. Marching two miles to Taben-Rodt to link up with B Company, the battalion found itself under artillery fire around 6:15 PM. Two rounds landed on Company A, resulting in twenty-four Ranger casualties.

Departing the town at 10 PM amidst sporadic and harassing German artillery fires, the battalion crossed a river on a footbridge and assembled a half mile northeast of the crossing site at 10:30 PM, still within friendly lines. Unfortunately, several more Rangers had become casualties of artillery fire during the march.

The Ranger Infantry Battalion crossed the American lines at 11:45 PM and moved forward of the 94th Infantry Division's sector. Though their route would take them through the Waldgut Hundscheid forest, the Rangers were not as concealed as they wished. German forests characteristically are crisscrossed by numerous trails that are seldom more than a few hundred yards apart. Despite the concealment provided by the trees and despite it being nighttime, the Rangers stood a good chance of being detected by large numbers of enemy patrols in the area.

As the lead company closed on the first checkpoint nearly two miles southwest of the objective, it encountered light small arms and machine gun fire. Several Rangers fell wounded. This engagement would be the first of four skirmishes that would mark the infiltration phase of the 5th Rangers' movement.

At 1:45 AM, the battalion's strength was further reduced when two B Company platoons became separated from the main body. Halting at a checkpoint around 6:30 AM, and establishing a perimeter defense, Lieutenant Colonel Sullivan waited for the two platoons to close with the formation. The wait didn't help, for the only group showing up was an enemy patrol. Capturing several of the enemy soldiers after a brief encounter, Sullivan quickly sent a reconnaissance patrol to find a safe route out of the area to the northeast. The patrol returned at 8:15 AM, and the battalion continued its advance with its prisoners in tow.

Despite the reconnaissance, the battalion was once again subjected to small arms fire from small groups of Germans. Ignoring the fire, the Rangers came within sight of the north edge of the Waldgut Hundscheid around 9:30 AM, approximately a mile southwest of the objective. After a brief halt, the unit continued its movement to the northeast. Five hundred yards later, Company D encountered stiff resistance. The company proceeded to outflank the defending Germans with one of its platoons and, following a short firefight, captured several prisoners. In the engagement, the Rangers suffered two more WIAs.

Though it was possible to continue to the objective along a concealed route through the woods, Sullivan elected to avoid the forest route and instead proceed east to cross a half-mile wide clearing to the north. Having been engaged during the infiltration on four occasions, keeping the men hidden was the least of his worries at the moment. The open route was shorter and would save time.

Company A led the battalion across the open expanse. Knowledge of the Rangers' presence in the sector appeared to be sorely lacking, for a German artillery officer, a medical officer, and several enlisted soldiers casually drove into the area and were immediately captured.

As they advanced, the Rangers methodically checked a cluster of houses near the center of the clearing and found them unoccupied.

As the rear unit, B Company, crossed the field, machine gun and small arms fire erupted from a medium-sized farm dwelling (called Kalfertshaus) with a pillbox nearby at the edge of the woods. A patrol was dispatched to deal with the menace. It accomplished the mission but, unfortunately, suffered one KIA. Occupying and securing the houses located in the clearing and having captured twenty-three prisoners, the battalion set about preparing to spend the night just south of the objective.

At 11:40 PM that same evening, just as the Rangers were settling in for the night, Task Force Riley, a subordinate unit of Combat Command B (CCB), 94th Infantry Division, passed through the 94th Infantry Division, captured Irsch, and continued its drive toward Zerf against strong German resistance.

The Rangers began their final push to the objective at 6 AM on 25 February. Confronting only light resistance and capturing more prisoners, the battalion closed on the objective by 8:30 AM and began to prepare its defensive positions. The location was not a good defensive position and presented some serious problems. The blocking position was set in a narrow strip of woods that extended north from the Waldgut Hundscheid forest and intersected the Irsch-Zerf road, which created a cleft in the wood line. Toward the west, the road sloped away approximately thirty yards to the front of the wood line positions the Rangers had established for cover and concealment. To the northeast and south stretched woods that provided a tempting avenue of approach for an enemy dismounted attack. Only the position facing east, toward the town of Zerf, provided suitable fields of fire.

Having no other choice than to make the best of the situation, Sullivan set about establishing his perimeter. E Company, located to the north, was responsible for placing antitank mines on the Irsch-Zerf road and covering them with fire. D and F Companies secured the east side, facing Zerf. Company C secured the west side. Company A was set up on the south side, where it could keep the Kalfertshaus under observation. In that B Company was still missing its two platoons, the remaining members guarded the prisoners in a barn near the center of the perimeter.

Apparently still unaware of the proximity of the Rangers, German elements continued traversing the road. Shortly after 12 PM, E Company captured a tank destroyer and destroyed it on the road with a bazooka to serve as a roadblock. Later a halftrack—a light armored vehicle—was destroyed by one of the antitank mines located on the road, and a group of German walking wounded were taken prisoner.

Organized German attacks on the Rangers did not begin until around 3:45 PM when Task Force Riley of CCB captured Irsch, and continued its drive toward Zerf against strong German forces entering Zerf from the north. Immediately following an intense artillery bombardment on the Ranger positions the Germans launched a two-pronged dismounted infantry attack along the covered and concealed routes offered by the forest. Company A was attacked by about two hundred infantry from the south while Company E had to face nearly four hundred infantry from the northeast. Though both assaults were repelled, it proved necessary to reinforce Company E with a platoon from Company F. The company commander and sixteen Rangers of Company B who had been guarding the prisoners filled the gap left in Company F's line.

An interrogation of eight enemy POWs identified the attacking unit as the 136th Infantry Regiment, 2nd Mountain Division, which had fought in the Arctic Circle on the East Front from 1941 through 1944. Having suffered heavily throughout the war, the regiment was seriously battle-fatigued. Though ordered to continue attacking the Rangers "to the last man," many of the 136th's soldiers chose capture to death. The POW total by the end of the day was 135 captives.

Having expended large quantities of ammunition during the infiltration and defense of the objective, the Rangers found themselves running low. At 4:20 PM, an attempt was made to resupply the battalion by air, but German antiaircraft weapons in the area forced the resupply plane to maintain an altitude of 1,500 feet above ground level. Hence, the majority of the parachuted containers landed outside of the Rangers' defensive perimeter. Without re-supply or relief soon, the battalion faced the possibility of running out of ammunition, and of being overrun.

During the night of 25 February, the Rangers continued to improve their positions. At 3 AM the next morning, a German force estimated at four hundred strong, attacked E Company's position. Kampfgruppe (Task Force) Kuppitsch was an assortment of miscellaneous administrative units, convalescent companies, and new recruits. Lacking unit cohesion, training, and any weapons heavier than rifles and machine guns, the motley collection of Germans compensated for the poor firepower with spirit and intense indirect fire support.

Hard-pressed, Company E was reinforced by the A Company commander and twelve of his Rangers. Accepting heavy losses, the Germans continued to apply the pressure, forcing the Rangers 100 yards from their positions. Only after the Americans called artillery in on their own overrun positions did the German force disengage. While capturing twenty-five more German soldiers, the Ranger head count indicated fourteen Rangers missing and believed captured.

Throughout the remainder of the early morning hours, intense artillery fire fell on the Rangers' positions. As the morning wore on, the intensity decreased and finally stopped altogether. A quiet descended upon the battlefield. At 11:55 AM on the 26th of February, Combat Command B reached the Rangers as the American mechanized force was driving on Zerf from the west. Combat Command B was able to meet the Rangers' most immediate needs: ammunition, water, food, and radio batteries.

Having evacuated the wounded Rangers and the seventy-five enemy prisoners, CCB continued its movement eastward while the Rangers remained in their blocking position throughout the night. A dense fog formed the next morning, resulting in an approach of a disoriented group of two hundred Germans. Seizing the initiative, the Rangers caught the enemy in a hasty ambush. Many of the Germans were killed. Realizing the futility of continuing the fight—being caught in a kill zone—the remaining one hundred and forty-five survivors threw down their weapons and surrendered.

Later around midday, the battalion was joined by the two lost platoons of B Company. Having returned to American lines after their break in contact, the two Ranger platoons had joined an armor unit and fought as conventional infantry alongside its tanks during the push to Zerf.

The final Ranger casualties were inflicted on the battalion by artillery fire that fell sporadically on the unit until mid-afternoon. Finally, at 3 PM on 27 February, nearly four days after crossing the line of departure (LD), the 94th Infantry Division relieved the 5th Ranger Infantry Battalion.

OBSERVATION

This action by the 5th Ranger Infantry Battalion in the vicinity of Zerf, Germany, was one of the most successful Ranger operations of the war. While infiltrating a battalion-sized element through enemy territory, securing the objective, and fighting off repeated attacks, the 5th Rangers killed 299 German soldiers and took 328 prisoners during an operation that lasted four days. The Rangers suffered approximately ninety casualties. The 5th's performance to move so stealthily over such a long distance and to set up a roadblock surreptitiously on the major road artery in the area significantly contributed to the collapse of the German front west of Zerf.

There is one additional note. Aside from the overwhelming success of this mission, the Zerf raid serves to demonstrate that much of warfare is not all action and glory. It is, in reality, a mosaic of overwhelmingly mundane, routine and dull events—"sheer boredom punctuated with moments of sheer terror" as described by many veterans. There are many forms of endurance and while physical endurance is impressive, so, too, is mental endurance when one trudges through the overwhelmingly physiologically and mentally exhausting experience as exemplified by this mission to achieve success.

Rangers move down a road as the lead of an attack on Heimbach. (3 Mar. 1945)

Rangers prepare to conduct an attack in Germany, vicinity of Heimbach. (3 Mar. 1945)

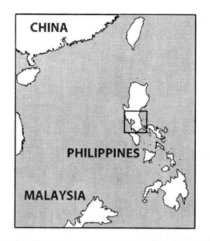

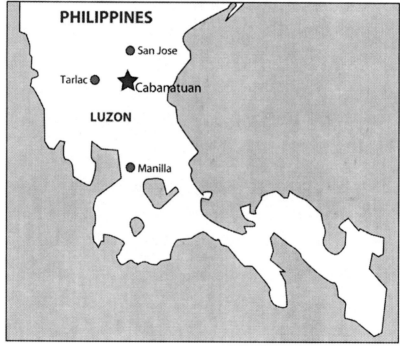

TEN

DARING

He who would greatly achieve must greatly dare for brilliant victory is only achieved at the risk of disastrous defeat.

—Washington Irving
(American author, essayist, called the
"Father of the American short story," 1783–1859)

DATE: *27–31 January 1945*

LOCATION: *Cabanatuan, Luzon, Philippines - Pacific Theater of Operations*

MISSION: *Liberate American and Allied prisoners of war from a Japanese prison camp located thirty miles behind enemy lines.*

BACKGROUND

SHORTLY AFTER TAKING COMMAND of the Sixth Army in 1943, Lieutenant General Walter Krueger created an elite force, patterned after the Navy's Underwater Demolition Teams (UDT) that conducted reconnaissance and special missions behind enemy lines. Designated the Alamo Scouts, these small teams of one officer and six enlisted men garnered nineteen Silver Stars, eighteen Bronze Stars, and four Soldier's Medals without suffering a single casualty during its initial nine-month period.

Significantly impressed, Krueger decided to create a larger force to accomplish the same type of missions as the Alamo Scouts—who would continue to remain in existence—but on a grander scale. On 25 September 1943, the 98th Field Artillery Battalion under the command of Lieutenant Colonel Henry A. Mucci was redesignated and reflagged as the 6th Ranger Infantry Battalion.

The 6th Ranger Infantry Battalion's initial baptism-by-fire was the invasion of the Philippines. Starting on 17 October 1944, three days prior to the actual landings on the main island of Luzon, the 6th Ranger Infantry Battalion landed on three strategic islands that guarded the entrance to Leyte Gulf with the mission to destroy Japanese communications facilities and weapons emplacements. Company D and part of Headquarters, Headquarters Company (HHC) landed on Sulvan Island, while the bulk of the battalion—those elements not engaged on the other two islands—landed on Dinagat Island. The next day, 18 October, B Company and the remaining elements of HHC landed on Homonhan Island. The 6th Ranger Infantry Battalion remained uncommitted thereafter for nearly three months until it landed on Luzon on 10 January 1945, and then only to spend the next two weeks guarding the Sixth Army's headquarters.

Its sojourn was about to end however. American and Allied POWs had been on Krueger's mind for some time. Aware of a POW camp at Pangatian, five miles east of Cabanatuan and thirty miles behind enemy lines, Krueger commenced planning the liberation of the camp as his army entered central Luzon.

Liberating the POW camp would be a significant challenge to anyone assigned the task. Deep behind enemy lines, the camp was situated on a major supply route (MSR) with Japanese units moving only at night along the route to avoid American airpower. Such movements would complicate the night activities of infiltrating units. Japanese tanks also used the roads in the area, and reports indicated a high concentration of Japanese troops in the nearby town of Cabu and Cabanatuan City. Additionally, the compound itself also served as a transit camp for units passing through.

In that the Japanese had already evacuated many of the prisoners from the camp, Sixth Army headquarters grew concerned for the welfare and safety of the remaining POWs, many of whom were survivors of the "Bataan Death March" and so emaciated and wasted away as to be referred to as "ghost soldiers." Fearing that the prisoners would be moved north or even worse, executed to prevent their liberation, Krueger assigned the difficult mission to the 6th Ranger Infantry Battalion.

THE LEGACY

For the planned mission, LTC Mucci assembled a team consisting of Company C, 2nd Platoon of Company F, two teams of Alamo Scouts, and four combat photographers. Under his personal command, the total strength of the assembled force was eight officers and 120 enlisted men.

The planning and preparation phase of the operation was exceptionally thorough. Aerial photographs in addition to map and ground reconnaissance provided the initial intelligence picture. Officers and enlisted soldiers studied the routes, rendezvous points, and location of the camp. The Army Air Corps would provide air cover. Sixth Army would provide intelligence updates through the bulky SCR 694 radio carried by the Rangers.

The Ranger uniform was fatigue-dress without rank or insignias and soft caps in-lieu-of helmets. Riflemen carried either the M-1 rifle or M-1 carbine, the weapons section, Browning Automatic Rifles (BARs), and most non-commissioned officers (NCOs), Thompson submachineguns and .45-caliber pistols. While Mucci was armed with only a .45-caliber pistol, most officers also carried a rifle. Besides his personal weapon, each man carried a minimum of two bandoleers of ammunition, a trench knife, and two hand or rifle grenades.

Both Alamo Scout teams departed the base camp on the afternoon of 27 January. They marched to a guerrilla headquarters at Guimba, where native guides led them to Platero three miles north of the objective. After linking up with the local guerrillas at that location, the Scouts kept the prisoner compound under surveillance, determining the number of Japanese guards, their routines, and the location of the prisoners. That intelligence would be furnished for the Rangers when they arrived.

Having completed their preliminary planning and waiting until they got closer to the target to complete the plan, Mucci's force finally departed its base camp at Calasiao at 5 AM on 28 January by truck. At Guimba they linked up with native guides and then force-marched five miles east to a guerrilla camp in the vicinity of Lobong. On arriving, a guerrilla force of eighty men led by Captain Eduardo Joson joined the Ranger force. Concerned about skirmishing with Communist Huk guerrillas operating in the area, Joson secured the camp with twenty of his men and sent the remainder out along Mucci's flanks to prevent the force from being ambushed during its movement.

Marching east through open grasslands and rice paddies, the 188-strong force crossed into enemy territory after dark about a mile south of Baloc. Fording the Talavera River around midnight and crossing the Rizal Highway around 4 AM, the force arrived at Balincarin at 6 AM on the 29th, having completed the fourteen-mile march from Lobong without a single incident or problem.

LTC Mucci linked up with the two Alamo Scout commanders, Lieutenants Thomas Rounsaville and William Nellist, who had departed from the base camp on January 27th only to learn that the Scouts were not finished with their reconnaissance. Shortly thereafter, the local guerrilla area commander, Captain Juan Pajota, and his force of approximately ninety armed and one hundred and sixty unarmed men joined the Rangers and Scout commanders in Balincarin. The size of the force was now 438.

The immediate situation was not good. Heavy concentrations of Japanese troops were in the area. Considerable traffic was passing in front of the camp, and two to three hundred enemy soldiers were bivouacked just a mile north of the compound on Cabu Creek. Also, there was a minimum of one Japanese division four miles to the south at Cabanatuan City. Convinced by intelligence that delay was prudent, Mucci postponed the raid for twenty-four hours.

The Company C commander, Captain Robert W. Prince, and the guerrilla leader, Captain Pajota, arranged for the guerrilla force to provide all-around security of the village. A convoy of carts pulled by water buffalos—carabao—large enough to carry two hundred liberated POWs was assembled. Additionally, they coordinated for the feeding of 650 men along the return route.

The guerrillas passed instructions to civilians within their area of responsibility. Chickens had to be penned and dogs tied and muzzled to prevent disturbances as the force moved by. Civilians north of the Cabanatuan City-Cabu highway were requested to detain any outsiders who wandered into their area until after the raid. And, for the safety of those living in the vicinity of the objective, they were told to leave at timed-intervals so as not to alert the Japanese.

Flanked by Joson's and Pajota's guerrillas for security, Mucci's Rangers departed Balincarin shortly after 4 PM for the two-and-a-half-mile trip south to Platero. The Alamo Scouts joined the force halfway to Platero to provide a situation report (SITREP). Verifying the previous reports, the Scouts added that another division-sized enemy force was heading toward Bongadon from the southwest. The decision to delay the assault had proven to be a wise one.

Platero's inhabitants warmly greeted the Rangers who entered the town at dusk. While most of the force rested, the officers and NCOs were planning the operation. They also transformed a one-story wooden building into a makeshift hospital. Meanwhile, the Alamo Scouts and guerrillas continued to reconnoiter.

The final reconnaissance report was very detailed. The objective compound was on the south side of the Cabanatuan City-Cabu highway. Measuring 600 by 800 yards, its perimeter was enclosed by three barbed wire fences, each separated by four-foot intervals and six to eight feet high. Less formidable barbed wire fences compartmentalized the compound. The prisoners appeared to be quartered in buildings in the northwest corner. An eight-foot-high gate secured with a heavy-duty lock barred the main entrance to the camp. A building inside the compound was believed to maintain four light combat tanks and two trucks. One sentry stood guard at the gate in a well-protected shelter. A team of four heavily armed soldiers occupied each of the three twelve-foot-high guard towers. Another four soldier team manned a heavily fortified pillbox. Seventy-three Japanese soldiers guarded the stockade. At eleven that morning, an additional one hundred and fifty soldiers entered the compound apparently to rest. Traffic on the nearby highway was light and the nearest outside threat came from the eight-hundred-man force supported by tanks and trucks at Cabu.

The attack would commence at dusk, and surprise was essential. Every member of the force was briefed as to his assigned mission. The Alamo Scouts had the camp under continual surveillance and used civilian runners to maintain communications between the Scouts and the main body, when necessary.

Mucci departed Platero at 5 PM on 30 January with a force numbering 375 men. Only the SCR 694 radio crew had been left behind with several armed villagers to provide security. Unbeknownst to Mucci, Pajota had a second force of four hundred armed guerrillas and four .30-caliber machine guns for additional support. Half of this force would be allocated to reinforce his blocking position; the other half was to serve as a reserve for Joson's roadblock. Pajota's rationale for not sharing this information was personal. He preferred to deploy these additional men according to his best judgment without having to discuss it with Mucci.

The force advanced for a half mile along a well-concealed, narrow dirt trail that pierced the tall grass and bamboo to the Pampanga River. At this point, the force split into three separate elements with Pajota and Joson leading their men across the river to their respective blocking positions.

Pajota's forces were to cut the phone lines linking the camp to the outside just prior to the attack and to establish a roadblock three hundred yards northeast of the compound at the highway bridge over the Cabu Creek for the purpose of stopping any reinforcements from getting to the camp from Cabu. Strengthened by the additional two hundred men he had sent ahead and his four .30-caliber machine guns, Pajota was able to cover the highway, the bridge, and likely river-crossing sites. Given that it was the dry season, the low water level of the creek provided a number of potential crossing sites. Additionally, an ambush was set up on the far side of the creek and a time bomb—one of several delivered by an American submarine— had been placed under the far end of the bridge and set to go off ten to twenty minutes after the assault started.

Joson's guerrillas, in the meantime, had moved eight hundred yards southwest of the compound to establish a roadblock on the main highway to stop any reaction forces from Cabanatuan City. Attached to the guerrillas was a six-man bazooka team from 2nd Platoon, Company F to provide anti-armor protection. Backing Joson up was the two-hundred-man reserve that Pajota had secretly provided.

After the departure of Pajota and Joson, Mucci led the main body across the river for the two-mile march to the objective. High grass concealed the first mile of its approach. At 6 PM, the main body broke from the tall grass to find a shrubless, treeless, flat, and barren rice paddy before it.

2nd Platoon, Company F, whose missions were to kill the guards at the rear of the compound, destroy a pillbox at the northeast corner of the compound, and prevent Japanese elements from moving into the prisoners' stockade, split from the main body and moved east. A half mile later, the platoon followed a stream bed that would conceal its movements to the east fence of the compound.

Company C's tasks were critical. 1st Platoon's mission was to breach the front gate of the compound and kill the Japanese guards in several known locations. Once that was accomplished, a platoon section would advance on the building housing the tanks and trucks and destroy it with bazooka fire. 2nd Platoon's mission was to follow the 1st Platoon through the breach to secure and to evacuate the prisoners.

Looking across the wide-open expanse before him, Prince led his two platoons forward another five hundred yards just to the point where the compound's watchtowers could be seen on the horizon. Assuming that "if he could see them, they probably could see us," Prince had his Rangers, the combat photographers, medics, the Alamo Scouts, and several guerrillas drop to the ground and begin crawling. It would take the Rangers seventy-five minutes to cover the distance.

At 6:40 PM, a P-61 Black Widow night fighter approached the compound. On Pajota's recommendations, the aircraft would provide a diversion, and possibly distract the Japanese guards from observing the Rangers' movement forward as they kept their eyes skyward. Flying over the bridge and prison camp twice at an altitude of two hundred feet, the fighter then departed the area on the prowl for enemy troops caught on the road.

Prince and his group completed their crawl approximately twenty-five minutes after Black Widow's departure from the area. Having arrived at a drainage ditch opposite the main gate of the compound and across the highway, the group waited for the assault to begin.

2nd Platoon, Company F, was still on the move while Company C waited in its attack position. Moving through a large culvert under the highway, the platoon advanced toward the back of the compound in a five-foot ditch that ran to the compound's east fence. At one point as the platoon was passing a guard tower, a sentry raised his weapon as if prepared to fire in the direction of the 2nd Platoon. Fortunately, the guard lowered his weapon and went about his business. The unit was finally in position by 7:25 PM.

In that the 2nd Platoon had the greatest distance to cover, Mucci selected the platoon leader, First Lieutenant John F. Murphy, to commence the assault on the objective. Though the attack was to be initiated at 7:30 PM, Murphy took some additional time to ensure his platoon was properly emplaced and prepared. Finally, at 7:45 PM, Murphy raised his rifle and, aiming at an open window of the nearest barracks, pulled the trigger. As darkness fell, the assault on the compound occupied by nearly 223 enemy soldiers had begun.

Small arms, automatic weapons, and grenades quickly followed. A lone sentry at the front gate was the target of so many weapons that his upper torso literally disintegrated under the hail of lead. Within thirty seconds, all pillboxes, guard shacks, and towers were "neutralized."

Staff Sergeant Theodore R. Richardson of Company C led the charge across the highway. Blowing the lock off the front gate with his .45-caliber pistol, he continued to advance into the compound. Bazooka sections moved to the central portion of camp and destroyed two trucks and a corrugated metal tank shed, but no tanks were found as had previously been reported.

As the Rangers were storming the compound, Pajota had his hands full at the bridge. With the Rangers' opening shots, the guerrillas opened fire on a Japanese battalion in bivouac approximately three hundred yards beyond Cabu Creek. The Japanese attempted to counterattack in a piecemeal fashion but were continually repelled with heavy casualties. The time bomb blew a gap in the bridge, and the four machine guns killed a significant number of Japanese as they attempted to cross the shallow river. Pajota's men were even able to destroy two Japanese tanks and one truck.

Twelve minutes after the opening round, enemy resistance came to a halt in the compound, and the Rangers began moving the American POWs out of the camp, carrying many of them on their backs. The Rangers suffered their first casualties of the operation when three light mortar rounds were fired in the vicinity of the front gate. Six men were wounded including Alamo Scout Rounsaville and the battalion surgeon, Captain James C. Fisher.

After a second sweep of the facilities for any stragglers, Prince fired one red flare into the air signaling the withdrawal. Unknown to Prince, a scared British civilian prisoner had hidden in the latrine during the assault but Filipino guerrillas fortunately rescued him later that evening. Tragically, another POW died of an apparent heart attack as he was being assisted out of the compound.

The last Rangers to withdraw from the objective were six men from Company F. Brought under fire, Corporal Roy Sweezy turned to return fire only to be hit in the chest with automatic weapons fire. He died a few minutes later.

By 8:30 PM the main body with its liberated American POWs had reached the Pampanga River. A total of forty-five minutes had elapsed. By 8:45 PM, these men were across the river. Prince fired the signal flare for the roadblock positions to withdraw. For Joson, that was relatively simple, for he had not fired a single shot in defense of his position. Undoubtedly, any Japanese plan to attack his position was checked when a P-61 fighter providing air cover for the operation strafed and destroyed a Japanese convoy at 8 PM heading toward Cabanatuan City and the roadblock from San Jose. Quickly withdrawing, Joson deployed half his men to Platero for local security while the remainder of his force provided flank security for the Ranger column as it left Platero.

Still battling a determined Japanese foe, Pajota was unable to withdraw as planned by the signal flare. The fight at the roadblock continued until 10 PM when the exhausted and seriously mauled Japancsc battalion ended its assaults against Pajota's position. At the cost of not even one serious casualty, Pajota's force had rendered nearly combat-ineffective an eight-hundred-man battalion. Conducting an orderly withdrawal, Pajota assumed the rear guard on the Pampanga River to prevent any pursuit of Mucci's column. All told, a total of 250 Japanese soldiers had been killed during the attack and defense of the road blocks.

The mission was not yet complete for the liberated POWs needed to be returned to friendly lines. The carts with the water buffaloes— carabaos—were waiting at the south bank of the Pampanga River as requested. The column moved to Platero where it stopped to reorganize, eat, and tend to some of the more seriously sick or wounded. Those ex-POWs who could walk departed for Balincarin at 9 PM under Ranger escort and were soon followed by the remainder of the column.

At Balincarin, more food and water as well as an additional fifteen carabao carts were added to the twenty-five already on hand. Unfortunately, the surgeon, Captain Fisher, died of his wounds in Balincarin. Corporal Roy Sweezy and Captain Fisher were the only Ranger KIAs of the operation.

The column departed Balincarin at midnight and moved to Matoas Na Kahey where it arrived at 2 AM on 31 January. Again, food and water and eleven more carabao carts provided relief. The column departed at 2:30 AM with fifty-one carts stretching a mile and a half long.

Just beyond Matoas Na Kahey lay the Rizal Highway. Not only did the column have to cross this dangerous area, but also because of the peculiar terrain features along the highway, the column had to enter the road at one point to the north and exit it at a point one mile to the south. Given the length of the column, two thirds of it would be exposed and vulnerable on the highway at one point.

1st Platoon, Company C, was given the task of providing crossing-site security. Armed with a bazooka and antitank grenades, one of the platoon's sections established a roadblock four hundred yards northeast of the column's entry point. A second, identical section, established another roadblock 3,000 yards to the south. The column took over an hour to clear the highway, amazingly completing the maneuver by 4:30 AM without being discovered.

The column halted for a rest stop at a small village around 5:30 AM and quickly moved on. Communications with the forward base at Guimba had been nonexistent throughout the operation despite the SCR 694 radio that had been lugged around. Repeated attempts to establish communications had failed.

At 8 AM, the column arrived at the small town of Sibul. Food and water were again provided with an additional twenty *carabao* carts thrown in for good measure. During the halt, the radio operators attempted radio contact with Guimba again. This time they succeeded and communications were established. Ambulances and trucks were requested to meet the column, which resumed its march to freedom shortly after 9 AM.

Technician 5 Patrick Marquis, on point and several hundred yards forward of the column, made contact with a Sixth Army reconnaissance patrol at about 11 AM. With the requested ambulances and trucks just a short distance away, the former POWs were at the 92nd Evacuation Hospital in Guima within an hour, mission accomplished for what would become known as the second most daring raid—second only to the exploits of the 2nd and 5th Ranger Infantry Battalions at Normandy—in U.S. military history. Later, as the rescued prisoners sailed back to the United States, tens of thousands of Americans turned out to cheer the men on as they passed below the Golden Gate Bridge.

Lieutenant Colonel Mucci was awarded the Distinguished Service Cross for planning, implementing, and leading the raid. The Silver Star was awarded to all American officers and the Bronze Star was awarded to all American enlisted men, all Filipino officers, and all Filipino enlisted men who participated in the raid. Let it be noted that the mission's success would have been highly unlikely were it not for the Filipino assistance, their friendship and support both advancing and returning from the POW camp objective.

OBSERVATION

The Cabanatuan prison camp raid was an overwhelming and stunning tactical success. The raid liberated 511 American and Allied POWs and resulted in an estimated 523 Japanese casualties at a cost of two Ranger KIA and ten WIA. General Douglas A. MacArthur stated that the raid was "magnificent and reflect[ed] extraordinary credit to all concerned."

Daring does not necessarily forego attention to detail as demonstrated by how meticulous the planning was for this raid. The coordination of the scouts, the use of the guerrillas to obtain transportation and food, the instructions issued to the local populace on where and when to move, and even the realization that animals needed to be penned and muzzled along the route of march to prevent disturbances are clear indicators that daring can be the successful melding of both physical courage and intellectual preparation.

The incorporation of guerrillas and the local populace into the plan for intelligence gathering, tactical reconnaissance, and security cannot be overstated. Often, commanders, consumed with the need for surprise and fearing compromise, dismiss this valuable asset. Numerous small incidents can cascade to disaster during a deep raid, as happened in Europe two months later when an armored task force of over 300 men and 50 plus vehicles met disaster in a similar raid at the POW camp in Hammelburg, Germany. The Cabanatuan raid succeeded in large measure from the assistance of the Filipino guerrillas and populace. Together, they added an intangible depth to the plan in case unforeseen actions of the enemy or just plain bad luck interceded. Dozens of likely incidents did not affect the raid because the Filipinos neutralized them early. History does not record them, but given Murphy's Law—where anything that can go wrong will go wrong, they are ever present.

*Rangers of the 6th Ranger Infantry Battalion head toward the
Cabanatuan prison camp on Luzon. (28 Jan. 1945).*

In column formation, the Rangers cross a shallow stream as they escort the released prisoners of war from the Cabanatuan prison camp back to friendly lines. (30 Jan. 1945).

Following their release from Cabanatuan by the Rangers, these former American prisoners of war rest on the porch of a temporary field hospital. (31 Jan. 1945)

Members of the 6th Ranger Infantry Battalion discuss their experience after having freed more than 500 American prisoners from the Japanese prisoner of war camp at Cabanatuan. One of the former inernees, wearing shorts, stands behind the group. (31 Jan. 1945)

Lieutenant Colonel Henry Mucci (left), commander 6th Ranger Infantry Battalion, greets Colonel James W. Ducksworth, senior officer among the ex-prisoners released from the Cabanatuan prison camp. (31 Jan. 1945)

ELEVEN

VALOR

No thought of flight,
None of retreat, no unbecoming deed
That argued fear; each on himself relied,
As only in his arm the moment lay
Of victory.

— **John Milton**
Paradise Lost, (English poet 1608–1674)

DATE: *25–26 November 1950*

LOCATION: *Hill 205, southeast of Unsan and 30 miles from the Yalu River, North Korea*

MISSION: *Defend a hilltop with an under-strength company against successive attacks of a 500-man enemy battalion during the Chinese Army's opening Thanksgiving Day offensive.*

BACKGROUND

THE KOREAN CONFLICT—OTHERWISE KNOWN as the "Forgotten War"—commenced on 25 June 1950 when seven Soviet-supplied and equipped North Korean divisions smashed headlong on the heels of a long and intensive artillery and mortar barrage into the totally unprepared units of the army of the Republic of Korea (ROK). Within two days, the capital of South Korea, Seoul, was abandoned and by the end of the first week a third of the ROK army was killed, captured, or missing. From its sedentary occupational army in Japan, the United States committed the 24th Infantry Division. Arriving in

Korea with 16,000 men on 1 July, the ill-prepared and poorly trained division was seriously mauled, losing 7,000 men by 22 July—an average of 318 men per day. Though serious losses had been inflicted on the attacking North Korean forces, the Allies were unable to stem the enemy's impetus southward, thus forcing the Allies to fall back and form a defensive perimeter around the southern port of Pusan under the command of the U.S. Eighth Army.

On 4 August, following a brief pause, the North Koreans assaulted what was known as the Pusan Perimeter. Along the northern perimeter, a small reentrant—forward position—between Taegu and Pohang-dong called the Pohang Pocket extended into the defensive lines. Throughout the initial stages of the conflict, U.S. Eighth Army leadership had been alarmed by the enemy's effectiveness with the infiltration of small, specialized units. Needing a special unit of its own to patrol the Pohang Pocket to provide timely information of the enemy buildup, the Eighth Army assigned the task of organizing a commando-style unit to its operations officer who passed the task along to Lieutenant Colonel John H. McGee, head of G3 Miscellaneous Division.

McGee had some experience in these matters. Having been captured and imprisoned in the Philippines during the Second World War, he managed to escape and to fight as a guerrilla against the Japanese. McGee's initial thoughts were of the Alamo Scouts, an organization he was familiar with from his earlier days in the Pacific Theater of World War II. A search of the archives for an Alamo Scout TO&E—Table of Organization and Equipment—to serve as the framework with which to build this new Eighth Army unit proved to be unsuccessful, however. Instead, the TO&E of a World War II Ranger Company was found and, thus, what would have been the Eighth Army Alamo Scouts became the Eighth Army Ranger Company.

On 8 August 1950, McGee began soliciting and interviewing potential candidates in Japan for a new unit. Based on a recommendation, the colonel sought out a 1949 West Point graduate, Second Lieutenant Ralph Puckett, Jr., 23, who was fresh out of Infantry Officer Basic Course (IOBC) and airborne school to command the unit, the Eighth Army Ranger Company (8ARC)—officially designated

8213 Army Unit. A team captain in boxing at West Point, the newly graduated lieutenant was always ready for a good challenge. As
Puckett recalled:

> I had just arrived at Camp Drake, Japan, the Replacement Depot
> for Korea. On my third day, I heard an announcement over the
> PA system telling me to report to Headquarters. I gave my most
> military salute and reported to the very distinguished lieuten-
> ant colonel sitting behind a folding table. Next to him was a first
> lieutenant.
>
> After giving me "At ease," Colonel McGee said, "I am se-
> lecting volunteers for an extremely dangerous mission behind
> enemy lines."
>
> I responded immediately, "Sir, I volunteer."
>
> He asked somewhat incredulously, "Don't you want to
> know what the mission is?"
>
> I answered, "Yes, Sir, but I volunteer."
>
> He said, "I am selecting volunteers for a Ranger Company
> soon to be formed for some very hazardous missions. I have al-
> ready selected the lieutenants."
>
> I said, "Sir, I want to be a Ranger so much that I volunteer
> to be a squad leader or rifleman if you will take me into that
> company."
>
> Colonel McGee talked to me at length. Before dismissing
> me, he said that he would make his decision and tell me tomor-
> row morning. When I reported to him the next day, he told me
> that he had accepted me and that I would be the Company Com-
> mander!
>
> I was amazed. I knew I did not have the experience to be a
> company commander. I probably didn't know enough to be a
> squad leader or rifleman!

The newly formed Ranger company of three officers and 73 enlisted men arrived in Korea on 2 September 1950—three lieutenants,
three sergeants, twenty-two corporals and the remainder privates
and private first class. The 1st Platoon Leader was Second Lieutenant

Charlie Bunn, 24—a West Point classmate of Puckett's. The 2nd Platoon Leader was Second Lieutenant Barney Cummings, 22—another West Point classmate of Puckett's, a National Collegiate Athletic Association (NCAA) fencing champion and the class "anchor" (currently called the "Goat" in today's West Point lexicon who is the last man in graduation order of merit for whom his classmates cheer loudest!). The company First Sergeant was Sergeant Charles L. Pitts, 24—an experienced soldier from World War II. The enlisted soldiers were all volunteers primarily from service units in Japan—cooks, mechanics, quartermaster repairmen. Few had any infantry background. Puckett had been told that experienced men could not be spared.

The men finally chosen were a breed of men who were not content to accept the monotony and boredom of a safe job in the rear. Instead, these new Rangers wanted action. They needed to be in the war, and they all volunteered to be in that war. However, they were nearly all "green" and virtually untrained in the hardships and rigors of infantry combat. Training, Ranger training, had to distinguish them if they were to survive. To achieve this objective, the new "shavetail"—untried Second Lieutenant, Puckett, had a plan.

We were certainly ordinary soldiers. Our training program had to build on the strengths that we had. We had to be molded into a team that was stronger than the sum of the individual soldiers that comprised that team. At the outset, I established four training goals:

1. Every Ranger would be in outstanding physical condition; every man would be a tiger!

2. Each man would be highly skilled in the tactics and techniques of the individual soldier.

3. Each squad, each platoon, and the company as a whole would be a smoothly functioning, highly efficient fighting machine.

4. Each Ranger would have the confidence in himself and his leaders that made him believe that he and the Eighth Army Ranger

Company were the best that our country could produce. This fourth objective was as important as any of the others.

Years later, the Eighth Army Ranger Company's training program of instruction has become an almost identical model by which The U.S. Army Ranger School trains men today. Standards are exceptionally high and perfection is always sought, especially by and within the chain of command. Individual and small unit fundamentals are stressed. Each tactical exercise is debriefed—a precursor of what is currently referred to as an "After Action Review" (AAR)—and lessons learned for improvement. Leadership positions are rotated while training, and the chain of command is emphasized with every man even to the most junior private receiving a full briefing on the mission and his personal role it. Physical training is an essential and integral aspect of everything they do. Most importantly, officers set the example, leading, participating and sharing the hardships together with their men. It is a proven formula that has withstood the test of time . . . and of combat.

Under the banner of the United Nations, the Allies launched a counteroffensive on 15 September. Planned by General of the Army Douglas MacArthur, a surprise amphibious invasion by the U.S. X Corps—the 1st Marine Division and the 7th Infantry Division—stormed ashore at the port of Inchon, 100 miles north and behind enemy lines. Within two weeks, the North Korean army was shattered and fled northwards as the U.S. 25th Infantry Division led a breakout from the Pusan Perimeter. A rapid push north across the 38th parallel—the border between North and South Korea—in early October soon opened the road through North Korea to the Yalu River, the border between North Korea and Communist China. By 26 October, elements of the ROK II Corps reached the Yalu. The total destruction of North Korea's military power—and thus an end to the conflict— seemed imminent.

Due to the urgent needs of the U.N. offensive, the Eighth Army Ranger Company conducted a five and a half-week course of instruction even though the original schedule called for seven weeks of preliminary training. Immediately after completing its initial Ranger training, the Eighth Army Ranger Company was assigned to combat duty on 11 October in support of the 25th Infantry Division, IX Corps, for anti-guerrilla operations. By then, the "dangerous mission behind enemy lines"—the "Pohang Pocket"—was overcome by events and, for the most part, it was believed by most that the need for "dangerous missions" would soon be over.

With a sense of final victory in the air, Puckett and his men were ordered to march to Kaesong in near-zero temperatures on the night of 18 November. The Ranger company was placed under the operational control (OPCON) of Lieutenant Colonel Welborn "Tom" G. Dolvin, commander of the 89th Medium Tank Battalion, who was forming a task force as part of the Eighth Army's final push north. Moved by truck, the Ranger Company finally linked up with the armored task force on 22 November.

During a meeting on Wake Island, General of the Army Douglas MacArthur had promised President Harry Truman that the boys would be home by Christmas. Towards that end, the Eighth Army prepared to commence the end-the-war offensive on November 24, Thanksgiving Day. The Chinese, however, were not idle. Throughout the months of October and early November, there were serious indicators of imminent Communist Chinese intervention. Two days prior to the offensive, on 22 November, the Task Force Dolvin Intelligence Staff Officer (S2) briefed to Puckett and other commanders that there were at least 25,000 Chinese in the 25th Division's area of operation. Armed Forces Radio announced that the Chinese had four corps between the Eighth Army and the Yalu River. Short-wave broadcasts were heard about so-called Chinese "volunteers" moving down from Manchuria to oppose the U.N. forces. Puckett immediately wondered, "Would the boys really be back home by next month?" Since there were significantly more than three Chinese to every American, the prospect of sending the guys home by Christmas suddenly seemed doubtful.

Thanksgiving Day dinner was served one day early. Traditional Thanksgiving dinners in the military are carved turkey with all the trimmings. For Puckett and his Rangers, the meal consisted of a cold turkey sandwich. *"C'est la guerre!"* — *"That's war!"*

Thanksgiving Day dawned clear with a temperature well below freezing. The Rangers were attached to Company B of the 89th Tank Battalion on the left flank of the task force. The Company B tank commander returned from the task force headquarters to inform the Ranger commander that his mission was to secure two hills miles to their front. Moreover, Puckett only had fifteen minutes to make plans and get them out to the Ranger platoon leaders before moving out. Puckett was prepared to move within the fifteen-minute limit, but he had his misgivings. He was not convinced that the tank company commander had adequately briefed his subordinates on the Ranger Company's scheme of maneuver or how they were to provide direct fire in support of the infantry attack. Though the Ranger commander knew this could be a problem, he had no choice but to have his men mount the tanks and move out.

The 25th Infantry Division's center zone of attack was to be spearheaded by Task Force Dolvin and would require the Rangers to execute a tough climb under hostile fire to sweep and clear the first objective hilltop, Hill 224 (some accounts refer to this as Hill 222). The advance began late afternoon from the Chong-Chon River. Traveling north along a dirt road, the combined tank and Ranger team began to take some machine gun fire approximately 100 yards from the objective hill. Puckett had his men jump off the tanks and form an assault line with the 1st Platoon on the left and Puckett with the 2nd Platoon on the right. As the Rangers began their assault, Puckett thought the tankers would begin to fire in support. Instead, they simply "buttoned up"—closed their hatches—and sat there silently, passively.

As Puckett would recall:

The bullets were zinging by, sounding like angry hornets. George Washington wrote his brother shortly after the former had been in a battle with the French in 1754. "I heard the bullcts whistle and, believe me, there is something charming in the sound." I never agreed with the words of my favorite hero. The crack! of bullets as they break the sound barrier has always been a terrible sound to me. I yelled at my men, "Let's go, Rangers!"

The ground was frozen, and only the stubble of rice stalks remained. There was no area to hide except behind small dikes throughout the paddy. The Rangers of the 2nd Platoon ran across the open rice paddies, taking fire as they did. Private First Class Joe C. Romero, 17, serving as his platoon's scout, led the platoon assault. As he spearheaded the platoon, he drew much of the enemy's firepower. Repeatedly exposing himself to these fires to determine enemy locations, he returned fire as he continued to advance up the hill. Quickly, Romero and some of his squad mates reached the top of the hill to find it unoccupied, the enemy having withdrawn before the determined attack. Below them, the remaining Rangers scaled the hill.

The enemy continued firing from a distance, the bullets hitting the dirt in front of Romero. As he began to scramble for cover, he was soon struck by machine gun fire, and though seriously wounded, refused to be evacuated and continued to place accurate fires on the enemy.

Suddenly, without warning, the lead tank that Puckett had ridden on finally opened fire. Four main-gun 76-mm rounds were fired in quick succession, impacting within ten feet of where Puckett stood and two feet from Sergeant Mackey D. McKinnon. Tank machine gun fire followed. McKinnon and Private First Class William J. Murphy were killed instantly by their own side. Sergeant Harry H. Cagley, Corporal Jesse E. Anderson and Private First Class Sadamu Tabata were wounded. Tabata, out of anger, fired several rounds from his M1 carbine rifle at the tank.

Puckett, incredibly unharmed by the tank fire, ran from the top of the hill down into the valley below to ensure that the tanks did not fire on his men again. The tank commander's failure to brief his subordinates had resulted in a tragic and deadly mistake. Meanwhile, the Rangers continued up the hill, clearing the entire rise. There they found Romero dead, shot through the chest. Cagley, though severely wounded by the friendly fire, continued to move about shouting encouragement and directing the evacuation of the wounded before he allowed anyone to help him.

With the dead and wounded attended to, the Rangers began to dig defensive positions into the frozen ground in anticipation of a counterattack. First Sergeant Pitts worked to alleviate the Ranger's anger and frustration by telling them "these things could be expected in war" and that "we must continue to do what we were called to do."

Meanwhile, Lieutenant Puckett was having his own words with the Tank Company commander. Amazingly, the tanker was criticizing the infantry lieutenant for not being with him in his tank command post. Puckett's angered reply was that the armor captain might be able to command his tank company in the rear of the action, but infantry commanders had to be with their men to lead them. For the Ranger company commander, it was a matter of priority and common sense.

The temperature plummeted to near zero. The Rangers' sleeping bags did not arrive, and the men suffered in the bitter cold. Frostbite was as deadly an enemy as the North Koreans and Communist Chinese. For many, the cold became so unbearable they had to build fires despite the danger associated with giving away their positions. Foxhole buddies Sergeant John Summers, 22, and Private First Class William L. Judy, 17, traded off keeping each other's feet warm. Recalled Summers:

> He [William Judy] would rub my feet and stick them under his arms. When my feet were warm, [he] took off one boot at a time and I'd warm up his the same way. This would not seem as much to some, but as cold as it was at that time, it helped to prevent frozen feet.

Sleep was out of the question in the frozen and brutal ground of the barren hill as the men persevered through piercing cold and the uncertainties of an enemy attack. First Sergeant Pitts transported five of his Rangers to the task force's aid station to be treated for frostbite.

When the sun rose the following morning, the Rangers felt as though they'd been given a second chance on life.

THE LEGACY

The morning of 25 November brought another cold, clear day. The previous day's action and the night's arctic temperature had taken their toll, though. Only fifty-one Rangers and nine attached Republic of Korea (ROK) soldiers were present for duty, about a reinforced platoon in number.

The task force resumed its attack at 10 AM with the Rangers once again riding the tanks into battle. The mission was nearly identical to that of the day before, except this time the objective was Hill 205 and known as "Objective 8" to Task Force Dolvin. Hill 205, thinly-wooded with pine trees, rose two hundred and five meters—615 feet—above the Chongehon River, southeast of Unsan and 30 miles from the Yalu River. The real bad news was that Hill 205 was about one to one and a half miles from the nearest ground unit support.

Around 2:30 PM, the mounted task force of Rangers encountered enemy resistance from Hill 205 as it approached, taking fire on the right flank. Buffeted by mortar, machine gun, and small arms fire, the small force stopped approximately half a mile from the base of Hill 205. The tank commanders quickly halted their vehicles and buttoned up. The Rangers jumped off the tanks to seek cover behind a low dike to the front. Behind them, the tanks remained inactive as though abandoned, and not a shot was fired.

Puckett, exposed to enemy fire, ran to the rear of a tank and tried to open a field phone box to talk to the tank commander inside. But the box would not open. Frustrated and with his adrenaline pumping, the Ranger then jumped on the tank and began banging the butt of his rifle against the armored hatch. For some of Puckett's Rangers, this image is unforgettable.

Eventually, the hatch cracked open a few inches and the tank commander peered out. With some well-chosen words, Puckett let him know that he was supposed to be firing on the enemy and that he should stick his head out to see what was happening. When the tanker complained that he only had three inches of steel between him and the enemy and needed to keep the hatch shut, Puckett reminded the man in no uncertain terms that he and his Rangers had only one-sixteenth of an inch of field jacket material protecting them. Apparently convinced, the tank commander disappeared back down the hatch. Soon, there was a flash, bang and recoil of a main gun round being fired.

With that, Puckett jumped off the tank and tossed off his coat and hat. Yelling "Let's go, Rangers!" to his 1st Platoon, the Ranger company commander led the attack and began to run towards the hill. In the distance, the 2nd Platoon began to move out as well.

The company had a bit less than half a mile of open rice paddy to run across to get to the base of Hill 205. Within range, enemy small arms fire focused on Puckett who was in the lead. The Rangers quickly sought what cover they could find behind the little rice paddy dikes. One enemy machine gun nest was giving Puckett and two of his Rangers, Corporal Barney Cronin and Private First Class Harland Morrissey, a problem and threatened the success of the attack by keeping them pinned down. Unable to determine the weapon's location, Cronin tried to deceive the gunner into firing and revealing his location by placing his hat on the barrel of his weapon and waving it overhead. The enemy gunner was not fooled, and held his fire. The enemy position remained undetected.

Puckett realized that a live target was a temptation the enemy most likely would not resist. Without hesitation, he offered himself as the target. Directing Cronin to keep his eyes open, the Ranger company commander took a deep breath, jumped up, and sprinted across the open space. The enemy fired and missed as Puckett threw himself to ground. Neither Cronin nor anyone else could spot the weapon's location. Puckett jumped up again and retraced his steps. The enemy fired and missed again. Cronin still failed to spot the enemy's location.

"OK! This is my last time!" Puckett declared as he jumped up a third time and once again ran across the open paddy. This time, Cronin was able to spot the enemy gunner and silenced him with a heavy burst from his Browning Automatic Rifle (BAR).

That threat neutralized, Puckett began to pick himself up off the ground, yelling at his men, "Let's go!"

Behind him he could hear Morrissey echoing his command, "OK, crazy Puckett says: 'Let's go!'"

Puckett hesitated, then turned to Morrissey with a smile. "You're right. Let's go!"

Puckett and the 1st Platoon continued across the remaining distance to the base of Hill 205. Along the way, close air support (CAS) strafed and napalmed a hill to the right from which the Rangers were receiving enemy fire. A shower of spent caliber .50 shell casings cascaded down on the running Rangers as the CAS flew overhead, prompting a surprised Puckett to dive to the ground. The chagrined Ranger could only laugh along with his men as he picked himself up off the ground and continued to move.

At the base of Hill 205, the 2nd Platoon joined Puckett and the 1st Platoon. The attack had cost the company three seriously wounded Rangers—Sergeant Melvin L. Hoagland, Corporal John V. Dzurcanin, and Corporal Billy S. Landers.

With the order "Fix bayonets!" Puckett led the remaining forty-eight Rangers and six ROK soldiers up the hill to find the objective unoccupied by the time they reached the top. Reorganizing and consolidating on the objective with a 360-degree defensive perimeter, leaders sighted their crew-served weapons on likely enemy approaches. Meanwhile, others began to hack away at the frozen ground with their entrenching tools, realizing somewhat unsettlingly, they were digging in a small Korean burial ground.

Slowly, foxholes were scooped out. While digging, Sergeant Merle Simpson, 21, heard a shout and looked up to spot a soldier in a strange uniform running down the hill. Before anyone could realize that he was a Chinese soldier, who had mistakenly stumbled into the Ranger's position, the soldier had disappeared into the tree line below. Voices could be heard in the distance, and they were not Korean. From the hilltop, the Rangers could see Chinese soldiers in a little town below, sitting around smoking and waiting for night to fall. The Rangers dug faster for they all knew what would be coming.

Puckett walked the perimeter with Lieutenant Cummings until he was satisfied with the defensive plan. Then, leaving Cummings in command, Puckett returned to the task force command post. He was compelled to do so for the company's radio was only functioning intermittently and fire support needed coordination.

Drawing enemy fire again as he ran across the open rice paddy at the base of the hill, Puckett reached the platoon of tanks they had left earlier only to find one of his Rangers lying behind a small paddy dike. The man was unharmed. Lying there, the panicked man cried out in a quivering voice, "I just couldn't help myself. I couldn't make myself do it!" The Ranger commander spoke softly and quietly to the broken man as he tried to convey to the soldier where his professional duties lie. "We really need you on that hill. We're going to have a tough battle tonight and your men are going to need you. Come on up the hill." However, no degree of imploring or cajoling would budge the man. With no more time to waste, Puckett abandoned him to continue making his way to the battalion headquarters command post (CP).

At the CP, Puckett plotted the artillery concentrations on his map with Captain Gordon Sumner, the battalion Fire Support Coordinator. Puckett also took the time to evaluate the overall tactical situation as laid out on the battalion operations officer's (S3) map overlay. What he saw, he did not like: his company, sitting on the hill top, could easily be surrounded by the enemy. There was, moreover, a mile or so gap between the Rangers and the nearest task force units. Reports coming into the command post clearly indicated that enemy resistance was stiffening, and a battalion that had been held in reserve had already been moved and positioned to reinforce Task Force Dolvin. Already, an infantry platoon in the task force area of operation had been hit hard by a Chinese assault.

Breaking down his rifle and cleaning it in the dark, Puckett waited for Lieutenant Colonel Dolvin to return. After a brief meeting with Dolvin, the Ranger linked up with First Sergeant Pitts who was serving as the company's only rear echelon. Despite Pitts' protestations, Puckett ordered the senior NCO to remain behind. The company commander needed someone reliable to ensure his unit was fed and resupplied. The Rangers could not fight without "beans and bullets," and no one was more reliable than the First Sergeant at obtaining them.

Reluctantly accepting his order, Pitts drove the lieutenant back to the open field, about half a mile from the base of the hill. From there, Puckett, along with Private First Class Merrill Casner, age 18, who'd stayed behind after their attack on Hill 205 to care for the wounded, moved across the rice paddy lugging ammunition and hand grenades with them. In the darkness, there was no threat of enemy fire.

During Puckett's absence, his seriously depleted company had continued to dig in, preparing for the inevitable Chinese counterattack that it knew would take place later that night or early morning. The men were battle-weary and miserable as the temperature continued to drop. A piercing wind on the hilltop plus the chill factor caused the temperature to drop well below zero. Rangers shivered, frozen, as they hunkered down, no fire, no sleeping bags. At 9 PM, the Rangers listened to a firefight in the distance, not knowing that swarming Chinese forces had just overwhelmed a friendly platoon.

The weather was incredibly harsh. Days before during a night march, a Ranger had found four chickens that he'd stuffed into a sack with the intent to eat them later. By dawn of that day, they had frozen solid.

Puckett and Casner reached the company perimeter about 10 PM, yelling as they approached to ensure they were not shot by mistake. Just as the two Rangers entered their lines, Chinese bugles and whistles could be heard down the far side of the hill. From the tree line, loud speakers boomed stilted English. "Tonight you die, Yankees!" In exchange, the Rangers shouted obscenities. Puckett ran to his foxhole with Lieutenant Cummings and Corporal James L. Beatty, 28, (one of the two oldest Rangers in the company) to see if his radio functioned properly in preparation for calling in artillery fires.

Suddenly, a mortar barrage cascaded on the Rangers' positions, the opening of a sequence of brutal Chinese attacks. Lifting its mortar fires, the Chinese ground assault against the Rangers commenced with another blowing of whistles and the blaring of bugles. Swarming up the hill amid a storm of hand grenades, the Chinese attack was met by an overwhelming fusillade of small arms and grenade fire from the Rangers. Sparks showered from arming devices as Chinese grenades were hurled through the air. Seconds later, they exploded, spraying dirt, rock, and shrapnel in all directions. Ranger machine guns and carbines fired continuously at shadowy figures coming over the hill line in between the flashes of gunfire and exploding hand grenades.

From his company CP, the foxhole, Puckett called in the artillery support. The preplanned firing missions that Puckett had coordinated with Sumner at the task force CP began to decimate the assaulting Chinese formations. Artillery flare shells floated leisurely overhead, their glare casting ghostly shadows over the forsaken piece of ground. As they slowly drifted down suspended from their tiny parachutes, they exposed a second wave of Chinese following the lead assault squads. With the attackers visible, the Ranger small arms and machine gun fires intensified. Enemy bodies continued to pile up. Finally, the first assault having failed, the Chinese withdrew.

As Private First Class Judy described the first assault:

Very shortly after the barrage, they came up the hill directly to our front firing as they came. Being dark, I couldn't estimate how many but it was obvious to me that we were outnumbered. [Corporal Robert P.] Sarama and I started returning the fire, hitting a few of them and causing them to stop and take cover about 75–100 yards down. Then they tried to crawl up the ditch line. I started rolling hand grenades down the ditch at about 5 minute intervals. We could hear them talking. It seemed that every time one of our grenades would go off some of them would yell so we knew we were doing some damage. All of a sudden, three of them charged our position from different angles close in. I got the guy in the middle with one shot. Sarama got the one on the left and the last [one] was coming from my right, firing his burp gun. At that inopportune moment, my rifle jammed so I yelled to Sarama, "Get this guy!" He swung his rifle over my head and fired, striking him in the upper torso because it actually turned him flip-flop and silenced him. Miraculously, neither of us had even been scratched. I cleaned my rifle and cleared the jam.

By 11:50 PM, the Ranger company commander reported to the task force command post that the attack had failed; the Rangers still held the hill. The attack was not without cost, though. Additional Rangers had fallen in defense of Hill 205. Leaving the safety of his own foxhole to run from foxhole to foxhole, checking the perimeter, and encouraging his men throughout the battle, Puckett was wounded by grenade shrapnel that had pierced his thigh. After the fact, Puckett had found the grenade incident to be an "amusing" one. Spotting the hand grenade trailing sparks as it sailed through the air towards him, the Ranger had immediately dropped to the ground as taught by The Infantry School at Fort Benning, Georgia. He knew from that instruction that the explosion and subsequent fragments would fly above his prone body. Upon detonation, however, he was literally shocked and almost scared to death to feel a fragmented shard cut into his thigh. His first thought was one of disbelief. "Benning was wrong?" Quickly getting back up, he sprinted to his foxhole as fast as his wounded leg would allow him where he shared his newfound revelation with his classmate, Cummings. "We both began to laugh as I realized how silly it all seemed."

Below the hill and across the rice paddy from where he'd dropped off Puckett, First Sergeant Pitts heard the small arms and mortar fire of the first assault. Moving back to the inactive tank platoon, he questioned the platoon leader as to what was happening. In the background, over the tank's radio, he could hear Lieutenant Puckett joking and laughing about being hit. Despite his wound, Lieutenant Puckett refused evacuation.

By now, the Rangers were not the only ones being assaulted. Allied units along the entire Eighth Army front in the west and the X Corps front in the east were under heavy attack. The big push to end the war and have the boys home in time for Christmas had not lasted very long, maybe a day or two. Instead, the entire U.N. command was now in a fight for its life. And, except for artillery support, the Rangers were in that fight alone. By their own efforts, they would either succeed or fail; their own training and perseverance would also help determine whether they lived or died.

Following the first assault, Puckett checked the perimeter again.

Crawling up to Judy's foxhole, he asked the corporal how he was doing. "OK, I guess; at least we're still here," was the reply. The lieutenant reached out and gave the Ranger a pat on the back and, after a moment, asked Judy if he was sure he was all right. Puckett held up his hand for Judy to see. It was covered with blood. "What happened? You get hit in the hand?" Judy asked. Puckett just shook his head. Judy then realized it was his blood. He hadn't yet felt the pain.

Judy's wound was the least of his concerns. He had another problem. Singling out a tree that was barely visible in the darkness twenty yards in front of them, he told Puckett that there was an enemy rifleman behind it who would pop out every couple of minutes to take a shot at him. Unfortunately, Judy was never able to get off a good shot in return.

Well, Puckett had been through this once before and felt up to testing his luck one more time. "Okay! I'm going to run across that open space. Maybe he'll take a shot at me and you can kill him. Be ready!"

Puckett began to sprint into the open. From behind the tree, the Chinese soldier appeared and took a shot at the exposed Ranger. Judy, in return, took a shot at the Chinese soldier. Both missed. From the ground, Puckett yelled out that he was going to try it again. Jumping up, he ran back across the open area. Again, with the same results: two shots; two misses. For the third time that night, the sixth in less than twelve hours, Puckett jumped up to serve as a human bull's-eye one last time. This time, Judy's bullet was on the mark, and the enemy gunner went down.

Throughout the next three hours, the Chinese launched four additional human wave attacks against the Ranger perimeter. While concentrated direct and indirect fires took their toll of the attacking enemy, there were moments when the Rangers' perimeter was breached as Rangers fell dead or wounded and weapons jammed in the near zero temperatures. Carbines locked up and machine guns were found to be frozen tight when triggers were pulled. The "spirit of the bayonet" was very much alive as Ranger bayonet assaults plugged the breaches.

Continuing to move about the perimeter checking the status of his men, interceding at each point of decision and helping to redistribute ammo—keeping only one eight-round clip for his own M1 Garand—Puckett steadied his men and directed his command's defenses. He warned his men to keep their heads down while calling in artillery fire "danger close" to place a high explosive wall of steel significantly closer to his positions than the by-the-book recommended 600 yards. To those Rangers he couldn't get to see personally, he'd yell encouragement to them, reinforcing the fact that they were in this fight together and that they could hold. Puckett knew that it was critical that his Rangers understood they were all in this together. That was key.

The intensity of the battle increased with each attack but throughout it all, Puckett and his Rangers remained calm and steadfast. As described later in 1989 by author Neil Sheehan in *Bright Shining Lie: John Paul Vann and America in Vietnam* (Winner of the Pulitzer Prize and the National Book Award):

Ralph Puckett had trained his cooks and clerk-typists well. They did not fall for the Chinese trick and cower in their foxholes from the blast of the grenades and the mortar shells. Instead, they raised their heads, picked out the figures running up at them through the night, and killed them as they came. Puckett helped them to aim and fling their own grenades down at the attackers by dispelling the darkness with a radio call to the artillery for flare shells. In the light of the flares, he could see more groups of Chinese soldiers running up the slope behind the lead squads. He dropped the next rounds of high explosive from the 105-mm and 155-mm howitzers right into them. Because Puckett was a conscientious lieutenant and because he had anticipated a fight further up the road, his men did not have to stint in their fire. He had made sure that every man was carrying a basic load of ammunition and then some to spare, and lots of grenades. The hilltop was a bedlam of carbines, rifles, BARs, and machine guns savaging the Chinese, while the volleys from the howitzers ripped them with shrapnel and tossed bodies into the air.

The sixth and final Chinese blow directed at Hill 205 was launched at approximately 2:35 AM as a battalion-sized force of over 500 men directed its main effort against Puckett's right perimeter held by the 2nd Platoon. A lengthy, intense mortar barrage initiated the assault. As the mortar fires lifted, bugles blew and the lead squads inundated the perimeter with hand grenades. Already significantly weakened by casualties and shortages in ammunition, the Rangers were unable to react quickly enough to this overwhelming threat. The enemy breached 2nd Platoon's sector and spread quickly everywhere as they swarmed over the platoon's position.

Having raced back to his foxhole from 1st Platoon's position upon hearing the bugles, Puckett vainly sought artillery support from the task force's artillery Liaison Officer, Captain Sumner. But the battery was firing in support of another heavily pressed unit at that moment and was unavailable. "We'll give you the fire as soon as we can."

On his knees, pressed against the wall of the foxhole, Puckett shouted into his radio's hand-mike to make himself heard over the din of battle as he struggled to convey the urgency of the situation. "We really need it now! We've just got to have it!"

By now, the Chinese were raining hand grenades everywhere, and mortar rounds were still falling. Then, two mortar rounds landed in the foxhole occupied by Puckett, Lieutenant Cummings and Corporal Beatty and exploded almost simultaneously. The foxhole was only six feet long, two feet wide and five feet deep. Mortar fragments ripped into Puckett's feet, buttocks, left shoulder and left arm. The damage to his right foot was so severe—severed toes and shattered bones—that he would later have great difficulty persuading the medics at the battalion aid station not to amputate it. Though seriously wounded and badly stunned, Puckett was still able to grasp the radio hand mike and plead one last time for artillery support, only to be told again that the guns were busy with another mission.

Without the artillery support, Puckett knew it was over and the battle lost.

"It's too late. Tell Colonel Dolvin we're being overwhelmed." The time entered in the task force's log was 2:45 AM.

With what little strength he had remaining, Puckett jumped from the hole and fell to the ground on his hands and knees after taking five steps. His shattered feet and wounds would not allow him to move any farther. Fifteen yards away, he could see Chinese soldiers bayoneting wounded Rangers. His company was being overrun. Wounds, exhaustion, and near zero temperatures had taken their toll, and he no longer had the strength even to raise his rifle in self-defense.

Out on the perimeter, Private First Class Judy was firing the last of his ammo. With enemy still coming up the hill and only an empty M1 rifle in hand, he hid behind a rock. The first enemy soldier who came around the rock caught the stock of Judy's weapon in the head, knocking the enemy combatant down the hill. The rifle continued its momentum and shattered against a rock. Now weaponless, the Ranger ran towards the platoon CP location only to stumble across the wounded Puckett who was still lying outside his foxhole. Feeling he would be nothing but a burden, Puckett ordered Judy to leave him behind. Wounded himself and realizing that his company commander was in no condition to move under his own power, Judy ran for help.

Entrenched in a foxhole on a finger of the hill running almost due north, Private First Class Billy G. Walls, 18, and Private First Class David L. Pollock were running low on ammunition for their BAR— Browning Automatic Rifle. In the distance, they spotted someone moving around the hill. Covering the man with their weapons as he approached, they soon realized that it was Judy.

"Our position has been overrun on the right," the wounded Ranger informed Walls and Pollock, "and Lieutenant Puckett's been seriously wounded." Pollock and Walls told Judy to locate their acting platoon sergeant and inform him of what had happened and to send assistance while they went to help their wounded commander.

Not knowing what they'd find, Pollock and Walls charged up the hill towards the crest. Ten yards from Puckett's location they cut down three enemy soldiers. While Pollock covered him, Walls knelt down next to the lieutenant. Nearby, a Chinese soldier was firing a burst of machine gun fire into a foxhole to finish off another Ranger.

"Are you hurt, sir? Can you walk?" Walls whispered.

"I'm hurt bad. I can't move. You'll have to leave me."

Rather than obey the command, the eighteen-year-old Walls handed his weapon to Pollock and began to lift Puckett off the ground. As he did so, the lieutenant directed Walls to check in the foxhole for Cummings and Beatty. Walls stopped to check the foxhole and the immediate area around it. Neither Cummings nor Beatty was anywhere to be seen.

"Sir, there's nobody in that foxhole," Walls reported.

With Puckett over his shoulder, Walls began to stagger down the hill as the Chinese began to close in on them. As the two Rangers slowly made their way towards a small draw thirty some yards away, Pollock stayed behind, holding off the advancing enemy with covering fire.

In the draw, Walls, exhausted, finally had to set Puckett down. "You're too heavy. I can't carry you any farther."

Soon thereafter, Pollock scampered into the draw.

"Is Lieutenant Puckett still with us?"

"I'm still with you. I'm not going to leave you," Puckett answered.

"We're not going to leave you, either, sir," was Pollock's reply. "Never will I leave a fallen comrade," the heart and soul of the Ranger Creed, already inculcated within the hearts and minds of Puckett's Rangers decades before that creed was even written.

Puckett, however, was concerned about leaving anyone behind. Walls and Pollock assured him that they were the last off the hill but each knew that was most likely not the case. The hill was overrun by the enemy, and there was nothing more they or their commander could do about that.

The sixth Chinese attack had finally overwhelmed the Rangers' position. But despite that, isolated Ranger elements continued to fight against staggering odds even as the enemy shot and bayoneted Rangers in their foxholes. Surviving the annihilation of his squad, Sergeant Simpson heard Judy yelling "Look out, Merle!" as the Ranger ran towards him. Turning, the sergeant watched as the Chinese started overrunning the remainder of his position. Together, the two Rangers ran towards the platoon CP, screaming a warning about the advancing Chinese along the way. Though only a Private First Class, Morrissey was serving as the de facto platoon sergeant. A World War II veteran who'd earned a Silver Star, the private was highly regarded by the Rangers for his experience.

Upon hearing the warning, Morrissey yelled, "Fix bayonets!"

"Fix bayonets?" Simpson incredulously replied. "There are hundreds of them!"

Looking about, Morrissey could see the shadowing outlines of the approaching enemy.

Realizing that discretion was the better form of valor, Morrissey begrudgingly agreed, "OK, let's go." Yelling for those around him to "get off the hill," Morrissey and the small band of Rangers began to move over, and then proceeded down the hill. A Ranger was hit in the arm but they continued to move. Bullets whizzed overhead as they ran and fell down the steep slope towards safety. To their rear, four brave Rangers—Corporal Sumner Kubinak, Private First Class Librado Luna, Private First Class Alvin Tadlock, and Private First Class Ernest Nowlin—sacrificed their lives by forming a rear guard to provide covering fires for the withdrawal.

Elsewhere, other Rangers made similar choices for their comrades in arms. Private First Class Harry Miyata, Private First Class Roger E. Hittle, and Private First Class Robert N. Jones also remained behind, sacrificing their lives while defending their positions and firing at the enemy with fierce determination as their comrades fell back.

Others, refusing to leave wounded comrades behind, heroically assisted their injured buddies off the hill as noted in the unit after action report.

> Moving about the fire-swept terrain, Sergeant John Diliberto organized his men for the withdrawal and started them on their way to more tenable positions. As he proceeded to fall back himself, he observed two of his comrades lying wounded on the exposed terrain. Without regard for his personal safety, he returned to the helpless men and dragged them both to safety as the enemy overwhelmed the defense perimeter.

Another of Puckett's men, Ranger Bill Kemmer, realizing that his Ranger buddy, Ted Jewell, was not at the bottom of the hill, returned to the top of Hill 205 and brought his wounded friend down the hill to safety.

Other Rangers were not so fortunate. Though overrun and overwhelmed, pockets of Rangers continued to resist, fighting to the death. Private First Class Casner, who accompanied Puckett up the hill earlier that night, lay seriously wounded by grenade fragments. Unable to move, he watched Private First Class Wilbert W. Clanton—a black soldier in an otherwise segregated Army—charge at a group of Chinese, yelling at the enemy, armed only with a bayonet in hand.

> The last Ranger I saw fighting on the early morning of November 26, 1950 at about [3:30 AM] was Wilbert Clanton. I will swear at least six Chinese were on him. How many he killed or wounded will never be known by us but I bet the North Koreans and Chinese know how many that Ranger took with him.

Casner himself had the muzzle of a rifle placed against his head and a shot fired. Fortunately for him, the resulting injury was not serious. After feigning death for the remainder of the night, he made his way back to friendly lines later that morning after the Chinese had departed the hill.

At the bottom of the hill, across the rice paddy, First Sergeant Pitts was gathering the survivors. Among them was Corporal Beatty who had shared the CP foxhole with Lieutenants Puckett and Cummings. Though blood covered his back, he was safe. Cummings, unfortunately, would be reported MIA and assumed killed in action KIA. When Pitts learned that the commander was still on the hill, he began to organize a rescue party with the twelve unwounded Rangers he could find.

Puckett, Walls, and Pollock still had life in them, though. Walls and Pollock had heard Morrissey yell "Fix Bayonets!" followed by his command to counterattack. Shortly thereafter, though, all had grown quiet, the calm enhanced by a full moon that almost made it appear to be day. Out of ammunition for their rifles, the weapons were broken down and discarded. All that remained was a single grenade. Puckett's thought was that while it wasn't enough to defend them, it was enough to ensure that they would not be taken alive. Above, they could hear the Chinese searching for survivors and starting to move over and down the hill. The three men needed to keep ahead of the enemy but, no matter how many ways they tried to carry the seriously wounded Ranger, they continued to move too slowly. Puckett told them to leave him. It was an order his men refused to obey. Finally, he told them to lay him down on his back and drag him unceremoniously by his arms.

For over two hundred yards they drug him over the ground. Wounded in three places, Puckett's pain was excruciating. Periodically, he would ask if they were the last ones off the hill to which Walls and Pollock responded, "Yes." Once, as they continued to drag him, Puckett began to mutter, reminding himself "I'm a Ranger! I'm a Ranger!" to fight the pain.

"Shut up, Sir, or you'll get us all killed!" admonished Walls and Pollock to quiet him.

Though the Chinese could not see them, they could certainly hear them as Puckett's body was dragged over rocks and branches, and the lieutenant's mantra of "I'm a Ranger!" was not helping. Bullets flew through the air and grenades were tossed in their direction as they continued down the slope. Though none of the men was hit, there were a number of close calls with Pollock's field jacket riddled with holes, especially its hood. Finally, as daylight began to break and about twenty yards from the tanks across the paddy field, they were challenged to identify themselves by some of the tankers guarding their perimeter.

Prior to the arrival of Puckett's three-man group, Morrissey, Judy and their small band of Rangers had safely made it down the hill. They had informed First Sergeant Pitts that Lieutenant Puckett was still alive. The First Sergeant called off the rescue attempt.

Along the tank unit's perimeter, the tankers were satisfied that Puckett, Walls and Pollock were "friendlies" and not Chinese. To assist the exhausted Walls and Pollock, a tanker sergeant came forward to carry Puckett back towards a tank. Still conscious despite the wounds and loss of blood, Puckett directed his Rangers where to place a tourniquet.

Still mindful of the situation around him, the Ranger lieutenant directed the tank commander to notify the task force headquarters that Hill 205 had been overrun and to place artillery fires directly on top of it. Shortly thereafter, a heavy concentration of 'Willie Pete'— white phosphorous—was fired on top of the hill. But the Chinese were not finished for some of them had worked their way down the hill and across the rice paddy where they began to fire on the tankers and the Ranger survivors.

Hurriedly, Puckett was tossed on top of a tank deck as everyone scrambled to get on board a tank. Loaded up, the vehicles quickly raced to the rear. At the Battalion CP, Puckett was debriefed by the Operations Officer. Shortly thereafter, he was finally taken to the battalion aid station. Later, back in the states at Ft. Benning, the fight to save his mangled right foot began in earnest.

Hospitalized because of the severity of his wounds until October 1951—nearly a full year—Puckett never again returned to command the Eighth Army Ranger Company. Instead, he went on to have one of the most distinguished careers in the United States Army. A Ranger legend and known to many as "The Ranger," Colonel Puckett retired from active duty in 1971 as one of the Army's most decorated warriors. The recipient of a second Distinguished Service Cross and two Silver Stars as a Battalion Commander in Vietnam, Ranger Puckett went on to not only become a charter inductee into the Ranger Hall of Fame in 1992, but also the Honorary Colonel of the 75th Ranger Regiment in 1996.

At the base of Hill 205 later that evening of the 26th of November 1950, First Sergeant Pitts assembled and reorganized the Ranger company survivors. As a testament to their tremendous sacrifice, only one commissioned officer and twenty-one enlisted soldiers of the original fifty-one who'd started to take Hill 205 the day before were present for duty by dawn of the 26th. Three of those twenty-two still fit for duty had not been on the hill during the battle, meaning that only nineteen of the forty-eight Rangers who fought, secured and defended Hill 205 were still combat effective—a numbing 61 percent casualty rate. Of those Rangers who were killed in action, the bodies of ten of them were never recovered.

In recognition of their heroic stand on Hill 205 against a force estimated to be nearly ten times their size, twenty of the fifty-two Rangers of the Eighth Army Ranger Company were awarded medals:

Lt. Ralph Puckett, Jr.	(WIA)	Distinguished Service Cross
Lt. Barnard Cummings	(MIA/KIA)	Silver Star
Sgt. John K. Kiliberto		Silver Star
Pvt. 1st Class Harland F. Morrissey		Silver Star
Pvt. 1st Class David L. Pollock		Silver Star
Pvt. 1st Class Billy G. Walls		Silver Star
Pvt. 1st Class Lucien J. Bourque	(MIA/KIA)	Bronze Star for Valor
Pvt. 1st Class Wilbert W. Clanton	(MIA/KIA)	Bronze Star for Valor
Crpl. Earle E. Cronin	(WIA)	Bronze Star for Valor
Pvt. 1st Class Roger Hittle	(MIA/KIA)	Bronze Star for Valor
Pvt. 1st Class Robert N. Jones	(MIA/KIA)	Bronze Star for Valor
Crpl. Sumner J. Kubinak	(MIA/KIA)	Bronze Star for Valor
Pvt. 1st Class Librado Luna	(MIA/KIA)	Bronze Star for Valor
Pvt. 1st Class Harry Y. Miyata	(MIA/KIA)	Bronze Star for Valor
Pvt. 1st Class Ernest G. Nowlin	(MIA/KIA)	Bronze Star for Valor

Pvt. 1st Class Alvin R. Tadlock	(MIA/KIA)	Bronze Star for Valor
Sgt. 1st Class John R. Ulery		Bronze Star for Valor
Crpl. Pierre Vaporis	(WIA)	Bronze Star for Valor
Pvt. 1st Class William J. Murphy		Army Commendation Medal for Valor
1st Sgt. Charles L. Pitts		Bronze Star

Although we failed to hold our position, I was proud of my Rangers. They had done exceedingly well against overwhelming odds. My Rangers had used what they had learned. They fought as they had been trained. I had trained my Rangers the way I had been trained in IOBC [Infantry Officer Basic Course]. They remembered the fundamentals. They were physically fit and, as a result, able to continue giving their all over a very long day and night. When the company was overwhelmed, many Rangers helped their wounded buddies off the hill. The Rangers deserve all the credit for the gallantry with which they fought that battle against overwhelming odds. They gave 100 percent and then some.

Ralph Puckett, Jr.

OBSERVATION

The blow that nearly annihilated the Eighth Army Ranger Company was part of a 400,000-man Chinese offensive initiated on Thanksgiving Day that surprised, staggered, mauled, overwhelmed, and broke United Nation units all along the front. Entire divisions were encircled and had to fight their way through Chinese roadblocks and ambushes established along main escape routes. The 25th Infantry Division was forced to fight a series of delaying actions as Chinese General Wu Xinquan's 39th Army pressed the attack. In order to rescue Task Force Dolvin, reserves under the command of the assistant division commander, Brigadier General Vennard Wilson, were committed thus saving the task force from destruction while also assisting the 25th's withdrawal to an area nine miles north of Kunu-ri.

Only when confronted with catastrophe, is it possible to truly test an army's resilience and cohesion. History is replete with defeated armies that were routed under such pressure. That the United Nations forces did not break is largely due to the valor of individuals, who in countless and often unrecognized cases, committed singular acts of valor that kept enough of the force intact to recover and stabilize the situation. In this sense, valor is much more than a willingness to sacrifice for one's immediate comrades. Many cumulative acts of valor can save an army. It's a given in the mind of every Ranger.

Beyond Lieutenant Puckett's compelling, "Follow Me!" leadership and the overall valor of the Rangers under his command, there are three other dynamic and fundamental lessons learned from the Eighth Army Ranger Company's courageous stand on Hill 205 that stand out from the many lessons that could be learned. The first is the necessity to establish and maintain realistic and demanding training standards. There are two maxims that every good combat leader adheres to when it comes to training: train as you'll fight, and sweat today minimizes blood tomorrow. Lieutenant Puckett created a training regimen that codified those maxims. It was a regimen that provided the confidence and comradery to stand against overwhelming and unrealistic odds that day in Korea, and it is still the regimen by which all Rangers are trained today.

The second lesson is this: keep everyone informed. If leadership and soldiers fail at this, there will be unnecessary loss of life or even mission failure. That was the case when the tank company commander apparently failed to brief his platoon leaders adequately, resulting in the unfortunate fratricide on Thanksgiving Day when a tank mistakenly fired on Puckett's Rangers as they secured Hill 224. The result was two KIA and three WIA. Keep your men informed, always.

The third lesson is that no matter how much training one does, no matter how well one's soldiers are prepared and informed of what's to come, there are those who will freeze or under-perform when that singular, telling and defining moment in their lives arrives. There is a "dark side of command," a side where men are sent to their deaths, a side where men knowingly go to their deaths, and a side where some men cannot respond to the call, no matter how hard they are pressed, no matter how logical the argument, no matter how much their comrades need them. Fear of dying or injury trumps all else.

For Lieutenant Puckett, that moment came when he stumbled across one of his Rangers who had grown too frightful of his own personal safety as he made his way from Hill 205 to the task force CP.

> This incident demonstrated a failure on my part to instill a will to "close with and defeat the enemy." As I have reflected on this incident many times in subsequent years, I castigated myself for not having focused on trying to prepare my Rangers for the emotional impact—the fear . . . even terror, confusion, self-doubt, pain from wounds—of combat. Why did I not spend time in training and before each attack talking to my Rangers with the one objective of allaying their concerns? I had never thought about it. I had just assumed . . . always a mistake . . . that all my Rangers would go about their responsibilities with complete professionalism. Had I overlooked the fact that all of us are human? That all of us have our own doubts, fears, dreams? Had I become so engrossed in "Accomplish the mission" that I had neglected "Take care of the men?" In my mind, those two injunctions are inseparable. You must accomplish both.

While Colonel (Retired) Puckett is certainly correct about the need to prepare Rangers—or any soldier—for the emotional impact of combat, there is also the reality that all are not necessarily fit for such combat and will fail when their courage deserts them. And, that need not be a demonstrative failure on the part of any leader.

There is one unique and final aspect to this story. Aside from the extraordinary valor displayed by the Rangers of the Eighth Army Ranger Company, it is the composition of the company, itself, that most astonishes. As noted in the background, with few exceptions all of these men were inexperienced, administrative soldiers with little or no infantry or combat-related experience. Their officers were three recently graduated lieutenants from West Point. The next ranking soldiers were three buck sergeants. The remainder were all corporals or privates. The oldest man in the company was in his early twenties; the youngest were sixteen—sixteen—years of age! The majority were seventeen to eighteen years old. No senior ranks, no infantry or combat experience and many were not even old enough to order a drink at a bar or old enough to drive, or even to vote. Now there is an incredible American story of valor.

Colonel Ralph Puckett, Jr.

Colonel Puckett was one of the most highly decorated soldiers in U.S. Army history [two Distinguished Service Crosses, two Silver Stars, three Legions of Merit, two Bronze Stars, five Purple Hearts, ten Air Medals]. He was tasked as a new second lieutenant out of West Point at the start of the Korean conflict to stand up and train the first Ranger company to be activated since the end of the Second World War.

Infantrymen preparing for an early morning patrol into enemy territory. (17 Apr. 1951)

A squad of infantry move out of an assembly area to probe enemy territory. (17 Apr. 1951)

Infantrymen riding on armor return after a probing patrol into enemy territory. (17 Apr. 1951)

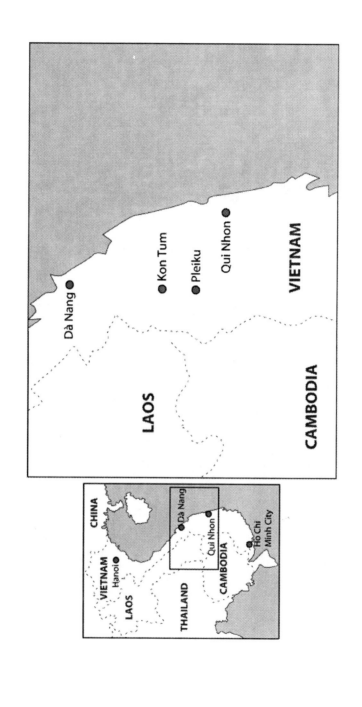

TWELVE

PERSONAL COURAGE

Courage is the first of the human qualities because it is the quality which guarantees all the others.

—**Sir Winston Churchill**
Second World War British Lord
of the Admiralty and Prime Minister, 1874–1965

DATE: *4–8 April 1970*

LOCATION: *Hill 763, Kontum Province, Republic of Vietnam*

MISSION: *American Ranger Advisor to an Army Republic of Vietnam (ARVN) Ranger Battalion assumes command to lead the defense against a North Vietnamese Regular Army Regiment attack.*

BACKGROUND

In 1960, COUNTER-GUERRILLA FORCES of South Vietnamese Army light infantry Ranger companies were created. Referred to as Biêt-Dông-Quân or BDQ, nearly 2,000 U.S. Ranger-trained officers and non-commissioned officers (NCOs) served as advisors to the Army of the Republic of Vietnam Ranger units throughout the duration of the war.

At its inception, the BDQ was composed of eighty-six companies. By 1965, they had been reorganized into battalions, three to five companies per battalion. Graduation from the South Vietnamese Ranger course was as gutsy as it gets—the real thing, a combat operation with only those returning alive designated as Ranger "graduates." By the end of March 1973 when the American advisors were phased out, there were a total of eight BDQ regimental sized groups with each regiment composed of two or three battalions each, for a total of twenty-two battalions. Each battalion had up to five U.S. Ranger officers and NCOs assigned to it.

Though they were supposed to operate only as advisors, some situations thrust the Ranger advisors into the forefront.

THE LEGACY

On 4 April 1970, Sergeant First Class Gary L. Littrell, a Light Weapons Infantry Advisor with Advisory Team 21, 11 Corps Advisory Group, found himself the only unwounded American Ranger advisor remaining after a surprise mortar barrage by the 28th North Vietnamese Regiment struck the 23rd ARVN Ranger Battalion, 2nd Ranger Group, in its defensive positions on top of Hill 763 in Kontum Province, where the battalion had established its base of operations. The mortar attack killed the South Vietnamese Ranger Battalion Commander, an American advisor, and seriously wounded all other American advisors other than Littrell. An attempted helicopter extraction of the wounded Americans led by Littrell failed as the helicopter pilots could not penetrate the heavy concentration of small arms fire to land near the Ranger NCO, who was standing in the open, exposed, holding a strobe light. After ensuring that the wounded were removed to a safe place, Littrell proceeded to direct close air support throughout the night and into the next morning.

At dawn there was another heavy mortar barrage against the besieged ARVN battalion. Littrell moved about the perimeter administering first aid, directing fires, and moving the growing number of dead and wounded. A resupply helicopter was able to make it into the landing zone (LZ) around 10 AM. Littrell loaded the three wounded American Ranger advisors on board along with some ARVN Rangers as another American Ranger, Specialist 5 Raymond Dieterle, disembarked from the helicopter with ammunition.

For the remainder of the day, the two Ranger advisors moved about the perimeter, calling in air strikes, adjusting artillery strikes, and encouraging their South Vietnamese Ranger compatriots. Enemy pressure against the perimeter grew all night long as sappers—combat engineers, demolitions experts—probed the battalion's defenses only to be driven back by artillery, aerial, and small arms fire. Under the circumstances, withdrawal at night, through woods and jungles, under heavy and punishing enemy pressure with a chain of command in disarray was not a viable option for the ARVN battalion. They would have to wait until daylight and the receipt of orders before they could attempt a withdrawal to safety.

The morning of 6 April opened with an attack by a North Vietnamese Army (NVA) unit that was repelled. Throughout the day, the NVA continued to rain heavy weapons and mortar fire down on the beleaguered force while the two American Rangers continued to make their rounds of the perimeter, bolstering Vietnamese's spirits. At 6:30 AM the following morning, another firestorm of enemy explosive volleys swept the hilltop. A massed assault staging in the woods at the base of the hill was met by helicopter strikes directed by Littrell at the point of attack. Despite the air strikes, the enemy's human-wave attack a half hour later nearly breached the tenuous perimeter. Only a determined defense by the fatigued defenders and the advisors' bravery repelled the charge.

The Rangers were able to maintain their position against a series of attacks that night. Finally, at 10:30 AM on 8 April, the depleted and exhausted battalion was ordered to withdraw down the hill, through the jungle, and across the Dak Poko River to link up with the 22nd ARVN Ranger Battalion. The two Ranger advisors set about organizing the movement, redistributing ammunition, and seeing that the dead and wounded were brought along.

The battalion moved out at 11 AM under the command of the dazed ARVN battalion executive officer who proceeded to move down the wrong ridge of the hill—despite the warning of the two American Rangers—and closer to the enemy position. At the bottom of the hill, the executive officer incredulously halted his formation for a five-minute tea break.

Immediately, the battalion was inundated with mortar fire. Littrell was able to establish radio contact with the 22nd ARVN Ranger Battalion and direct counter-fires against the enemy mortars. Requests for gunship support brought the word that none was immediately available, which prompted the battalion executive officer to panic and run. Seeing their commander flee, the remainder of the South Vietnamese Rangers began to scatter, leaving their dead and wounded behind.

With a "Come on, partner, let's hat up"—meaning let's get a move on it, Littrell grabbed Dieterle and began rounding up and organizing the dispersed ARVN Rangers. With the remnants of the battalion regrouped, Littrell and Dieterle proceeded to lead the formation through the jungle for several hours toward their objective—a secure location defended by the 22nd ARVN Ranger Battalion. As the North Vietnamese pursued, the American Rangers directed mortar fires "danger close"—calling in fires perilously close to them—to keep the enemy at bay. During the repulse of one assault, Littrell and Dieterle were knocked to the ground by the force of 500-pound bombs Littrell had called in nearby.

Directing the fight through two ambushes, the Americans continued to lead, cajole, and drag the ARVN soldiers along. At the last ambush site, Littrell stopped to assist three wounded Vietnamese Rangers. Carrying the most seriously wounded on his back, Littrell had to drag the other two behind him and across the Dak Poko River as they held on to his web gear—combat suspenders—and were led to safety by Littrell with the remainder of the battalion. Total losses for the ARVN Ranger Battalion were 218 casualties (KIA and WIA) and nineteen MIA, nearly 40 percent losses for a 600-man battalion.

OBSERVATION

In times of great peril, when a unit as a whole can begin to falter, the action of just a single man can make all the difference. Sergeant First Class Gary L. Littrell was just such a man. Though an American advisor in support of a foreign unit and not a member of the Army of Vietnam (ARVN) infantry battalion's chain of command, Littrell stepped forward in a time of dire need and took the initiative to establish a defense and call in supporting fires while motivating and directing the beleaguered Vietnamese command. Without question, this Ranger's actions prevented the total destruction of this battalion.

As a direct result of this incredible display of personal courage and for conduct above and beyond the call of duty, Sergeant First Class Gary L. Littrell was a recipient of the Medal of Honor.

Command Sergeant Major Gary Lee Littrell

MEDAL OF HONOR CITATION

LITTRELL, GARY LEE

Rank and organization: *Sergeant First Class, U.S. Army, Advisory Team 21, 11 Corps Advisory Group.*

Place and date: *Province, Republic of Vietnam, 4–8 April 1970.*

Entered service at: *Angeles, California*

Born: *October 1944, Henderson, Kentucky.*

Citation

For conspicuous gallantry and intrepidity in action at the risk of his life above and beyond the call of duty. Sfc. Littrell, U.S. Military Assistance Command, Vietnam, Advisory Team 21, distinguished himself while serving as a Light Weapons Infantry Advisor with the 23rd Battalion, 2nd Ranger Group, Republic of Vietnam Army, near Dak Seang.

After establishing a defensive perimeter on a hill on April 4, the battalion was subjected to an intense enemy mortar attack which killed the Vietnamese commander, 1 advisor, and seriously wounded all the advisors except Sfc. Littrell. During the ensuing four days, Sfc Littrell exhibited near superhuman endurance as he single-handedly bolstered the besieged battalion. Repeatedly abandoning positions of relative safety, he directed artillery and air support by day and marked the unit's location by night, despite the heavy, concentrated enemy fire. His dauntless will instilled in the men of the 23rd Battalion a deep desire to resist. Assault after assault was repulsed as the battalion responded to the extraordinary leadership and personal example exhibited by Sfc. Littrell as he continuously moved to those points most seriously threatened by the enemy, redistributed ammunition, strengthened faltering defenses, cared for the wounded and shouted encouragement to the Vietnamese in their own language. When the beleaguered battalion was finally ordered to withdraw, numerous ambushes were encountered. Sfc. Littrell repeatedly prevented widespread disorder by directing air strikes to within 50 meters of their position. Through his indomitable courage and complete disregard for his safety, he averted excessive loss of life and injury to the members of the battalion. The sustained extraordinary courage and selflessness displayed by Sfc. Littrell over an extended period of time were in keeping with the highest traditions of the military service and reflect great credit on him and the U.S. Army.

Members of Company C, 75th Ranger Battalion, move through to se-cure an open area near Da Lat, Tyuan Duc Province. (4 Mar. 70)

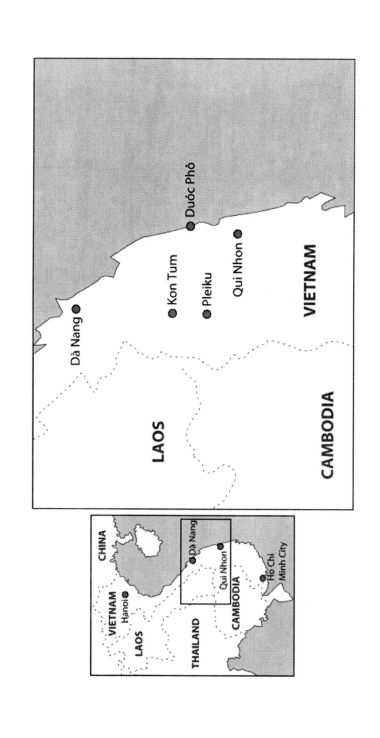

Thirteen

HEROISM

When the will defies fear, when duty throws the gauntlet down to fate, when honor scorns to compromise with death—this is heroism.

—Robert G. Ingersoll
(American orator and political speechmaker, 1833–1899)

DATE: *29 November 1969*

LOCATION: *Valley west of Duc Pho, Quang Ngai Province, Republic of Vietnam*

MISSION: *Conduct an ambush and capture an enemy prisoner.*

BACKGROUND

THE 197TH INFANTRY BRIGADE, 23rd Infantry Division—referred to by many as the "Americal Division"—arrived in Vietnam in August 1966 and was assigned to Quang Ngai Province in the southern part of Vietnam. In a support role was Company G (Ranger), 75th Infantry Regiment—the "Golf" Rangers, in reference to the phonetic pronunciation of the letter G.

On 18 March 1969, the regions formally assigned to the Americal Division and the 2nd Army of Vietnam (ARVN) Division were merged. The 23rd's area became a joint American-Vietnamese responsibility and Ranger Company G was directed to build two demonstration teams to show how U.S. and ARVN soldiers could work together. Though each of the two teams only had two ARVN Rangers assigned, they proved to be a valuable asset and were assigned lead and trail positions within the patrol. These were the most dangerous positions within the team, for the point and rear guard were the most likely positions to make enemy contact first. They were placed there for their native ability to discern friend from foe: these ARVN Rangers could tell the difference between a civilian and an enemy soldier just by the sound of their walking through the jungle; the difference being that civilians routinely plodded along, making more noise than a soldier who steps more carefully and stealthily.

The Americal Division Reconnaissance Zone was often referred to as the "Suicide Zone" for it extended from the interior mountains to the remote and isolated Laotian border. Having already had two garrisons destroyed by regimental-sized North Vietnamese Army (NVA) assaults, the U.S. Army Special Forces had even abandoned the region. Facing such adversity, the Americal Rangers limited reconnaissance missions within the area to twenty-four hours (between drop-off and retrieval). Longer missions would have risked compromise and destruction.

THE LEGACY

Staff Sergeant Robert J. Pruden commanded Ranger Team Oregon—a six-man team. On 14 November 1969, he and his team were conducting surveillance of a trail west of Duc Pho when they ambushed a ten-man Viet Cong patrol, killing a number of the enemy prior to withdrawing and being extracted from the area. Four days later, Pruden and his team—composed of Sergeant Danny L. Jacks, Specialist 5 John E. Shultz, Specialist 4 John S. Gaffney, Specialist 4 James R. Gromacki, and Specialist 4 Robert B. Kalway—volunteered to return to the same valley road to conduct another ambush and to capture an enemy prisoner.

Inserted on a hill by helicopter, the team moved into some rice fields and established two three-man positions on both sides of a well-used trail. Depending on the direction of travel of the intended prisoner, one team would make the capture while the other team provided rear security. As they were preparing the ambush site, a twelve-man Viet Cong (VC) patrol approached. Ceasing all movement, the Rangers watched quietly as the patrol moved through their ambush site.

With the patrol's departure, the preparations continued. Unfortunately, six VC spotted one of the Rangers as he crawled forward to position a claymore mine (a rectangular explosive device with 750 steel balls that would create what could be called a large shotgun blast) along the trail. The Ranger was pinned down by rifle fire from these six as the team began to take additional fires from an enemy force closing in from the opposite direction near a road bend.

Seeing one of his men cut off and exposed as the enemy closed in, Pruden realized that his team was in a precarious position. Leaping onto the trail from his concealed position, Pruden charged toward the road bend firing his CAR-15. Startled by this unexpected assault, the VC shifted their fires from the Ranger trapped on the trail to the Ranger charging toward them. Hit and knocked to the ground, Pruden rose and continued forward when he was hit a second time.

Falling along the edge of the trail, the Ranger paused long enough to slap in another magazine, having killed four of the enemy and wounding several others with his first clip. Rising to his feet once again, Pruden's boldness so unnerved the remaining VC that they fled. As they withdrew, though, they fired some departing shots, hitting the charging Ranger a third time.

In the meantime, having repulsed the second approaching enemy element while Pruden was attacking on the other end of the trail, the team closed in on its fallen leader to find him on the ground only fifteen feet from the enemy's previous position with five bullet wounds in the chest and abdomen. Amazingly, the fallen Ranger was still conscious. With a hasty defensive perimeter thrown around him by the team members, Pruden, despite the overwhelming pain, forced himself to remain conscious and even attempted to call in air support and an immediate extraction for his team, though he could barely speak above a whisper by this time.

As his breathing became more difficult, team members attempted to keep the sergeant resuscitated with mouth-to-mouth. Temporarily regaining some strength, Pruden asked the status of his team members and quietly reiterated his orders to the team that they were to be extracted. Moments later he lost consciousness and died just minutes before the arrival of the extraction helicopter.

OBSERVATION

A.W. and J.C. Hare, English clergymen and writers of the 17th Century once noted that "Heroism is the self-devotion of genius manifesting itself in action." Others note that heroism is a composite of various forms of courage that denote fearlessness and a defiance of danger with a fortitude that nobly bears up under trials, danger and suffering. Perhaps most fittingly, heroism can be defined as a disregard of danger, derived not from ignorance or thoughtless light-heartedness, but from a righteous devotion to some valiant cause, and an honest confidence of being able to meet that danger in the spirit of such a cause.

On 22 September 1776, a great American patriot and Revolutionary War hero, Captain Nathan Hale, epitomized heroism with the immortal words, "I only regret that I have but one life to lose for my country" before being hanged by the British as a spy at the age of twenty-one.

On 18 November 1969, such an epitaph was no less fitting for Staff Sergeant Robert J. Pruden whose gallantry, intrepidity, inspirational leadership, and personal self-sacrifice resulted in actions and an amazing display of heroism that led to the defeat of two flanking enemy squads and the safe return of his five team members. For conduct above and beyond the call of duty, Ranger Staff Sergeant Robert J. Pruden was posthumously awarded the Medal of Honor.

Staff Sgt. Robert J. Pruden

MEDAL OF HONOR CITATION

PRUDEN, ROBERT J. (Posthumous)

Rank and organization: *Staff Sergeant, U.S. Army, 75th Infantry, Americal Division.*

Place and date: *Quang Ngai Province, Republic of Vietnam, 29 November 1969.*

Entered service at: *Minneapolis, Minn.*

Born: *September 1949, St. Paul, Minn.*

Citation

For conspicuous gallantry and intrepidity in action at the risk of his life above and beyond the call of duty. S/Sgt. Pruden, Company G, distinguished himself while serving as a reconnaissance team leader during an ambush mission. The six-man team was inserted by helicopter into enemy controlled territory to establish an ambush position and to obtain information concerning enemy movements. As the team moved into the preplanned area, S/Sgt. Pruden deployed his men into two groups on the opposite sides of a well-used trail. As the groups were establishing their defensive positions, one member of the team was trapped in the open by the heavy fire from an enemy squad. Realizing that the ambush position had been compromised, S/Sgt. Pruden directed his team to open fire on the enemy force. Immediately, the team came under heavy fire from a second enemy element. S/Sgt. Pruden, with full knowledge of the extreme danger involved, left his concealed position and, firing as he ran, advanced toward the enemy to draw the hostile fire. He was seriously wounded twice but continued his attack until he fell for a third time, in front of the enemy positions. S/Sgt. Pruden's actions resulted in several enemy casualties and withdrawal of the remaining enemy force. Although grievously wounded, he directed his men into defensive positions and called for evacuation helicopters, which safely withdrew the members of the team. S/Sgt. Pruden's outstanding courage, selfless concern for the welfare of his men, and intrepidity in action at the cost of his life were in keeping with the highest traditions of the military service and reflect great credit upon himself, his unit, and the U.S. Army.

A typical Ranger LRRPs team. Picture is provided by Ranger Pat Ta-dina, pictured on the far left. Ranger Laszlo Rabel, another Medal of Honor recipient, is identified upper right. (1968)

SP4 Jim Messengill, Team 22, Co H (Ranger) 75th Infantry Regi-ment, reports contact with enemy forces during a patrol north east of Xuan Loc, along the Dong Nai River. (9–12 Dec. 1970)

SP4 Steven Paffel, Team 22, Co H (Ranger) 75th Infantry Regiment, prepares to load his M-79 grenade launcher after enemy contact was made during a patrol north east of Xuan Loc, along the Dong Nai River. (9–12 Dec. 1970)

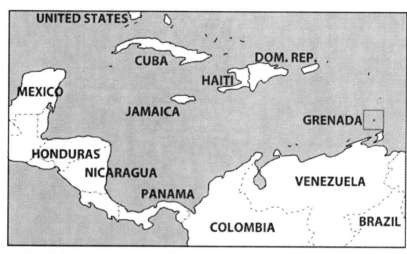

FOURTEEN

PLANNING

The plan was smooth on paper, only they forgot about the ravines.
—**Russian military proverb**

DATE: *24–25 October 1983*

LOCATION: *Point Salines Airfield, Grenada (Caribbean)*

MISSION: *Operation Urgent Fury, Airfield seizure and Non Combatant Evacuation Operation (NEO)*

BACKGROUND

ON 25 OCTOBER 1983, United States Army Rangers, as part of a seven-nation task force, led the invasion of the tiny Caribbean island of Grenada. Little was known about the enemy they faced, no tactical intelligence, or details of locations. Military maps of the island did not exist and much of the planning was based on an old, outdated British 1:50,000-tourist map of the region. Black and white photocopies of this map were distributed for planning purposes.

From both the 1st and 2nd Ranger Battalions, 600 Rangers would land or drop, depending on the conditions encountered at the airfield located on Point Salines. H-Hour (a military term that defines an hour reference point in the same manner that D-Day defines a day to signal the beginning of an operation) was set for 2 AM on 25 October. Lieutenant Colonel Wesley Taylor, commanded the 1st Battalion (Ranger), 75th Infantry Regiment, at Hunter Army Airfield (HAAF), Georgia, and Lieutenant Colonel Ralph Hagler, commanded the 2nd

Battalion (Ranger), 75th Infantry Regiment at Fort Lewis, Washington. Both officers seriously regretted that they would only be able to deploy with 50 percent of their unit's authorized strength for there were not enough Air Force crews trained for night-time C-130 operations to deploy the full two battalions.

2nd Battalion (Ranger) was to follow the 1st Battalion into Point Salines, assist with the securing of the airfield, and conduct a follow-on mission to attack Camp Calivigny. H-Hour was pushed back three hours to 5 AM, thirty minutes before dawn. The 2nd Battalion now had to drop and also conduct a field march in daylight in order to attack Calivigny.

THE LEGACY

The Air Force 1st Special Operations Wing would transport the Rangers to their drop zone. With two MC-130Es—aircraft that carry precision navigation and terrain-hugging equipment for deep-penetration and low-altitude night operations—leading the way, the 1st Battalion (Ranger) would be carried on five C-130s. The 2nd Battalion (Ranger) was also allocated five C-130s for their lift. Each MC-130E and C-130 carried sixty-plus Rangers each. Five AC-130H Spectre gunships from the 16th Special Operations Squadron out of Hurlburt Field, Florida, were assigned for fire support.

Leading the air armada was a group of Pathfinders. Free-fall specialists, the Pathfinders would parachute over Point Salines around 3:30 AM on the 25th from a reconnaissance C-130 aircraft flying overhead. Once on the ground, they were to radio the airborne Ranger Battalions with the status of the runway: would the Rangers be able to air land or would they have to jump? In addition, they would define the drop zone with lights. Their follow-on mission was to conduct a reconnaissance of the True Blue facilities—a medical school campus occupied by U.S. citizens at the eastern end of the runway.

An hour after the Pathfinders and thirty minutes prior to the arrival of the main Ranger force, A Company, 1st Battalion, loaded on the two lead MC-130s, would arrive to clear the air strip of enemy personnel and obstacles. The decision as to how to arrive—air land on the runway in the transport aircraft or parachute from them,

would be made at the last minute based on the best available information. Taylor and the remainder of the 1st Battalion in their five
C-130s would then drop and secure the Salines runway from Hardy
Bay, east. Following close behind would be Hagler's five C-130s to
drop and secure the Salines runway west of Hardy Bay.

During the evening of 24 October, as Taylor was having a final
word with his Rangers, the first seven MC-130s and C-130s arrived.
To his consternation, Taylor learned that the 1st Special Operations
and the Rangers' flight schedules differed by two hours and the 1st
Battalion—along with two heavy equipment operators of the 618th
Engineering Company from the 82nd Airborne—needed to load immediately. A frantic rush to load up ensued. Only twenty-five minutes late, the first MC-130 with A Company Rangers roared down the
runway at 9:30 PM. Even more annoying to the battalion commander
was the fact that the aircraft had arrived at Hunter Army Airfield
(HAAF) without hatch mounts—special fittings and openings in
external hatches, doors or windows—for antennas which meant he
could not communicate directly with his men while in flight. Consequently, Taylor had no option other than talking to his men through
the aircrew's radios, forcing the Rangers to make their way through
the crowded aircraft to get to the devices.

Cramped, the Ranger leaders huddled up front in the aircraft as
they reviewed their missions. In red nylon seats along the sides of
the aircraft, the Ranger rank and file waited, trying to find some comfort and get some sleep during the long flight packed like sardines
in a can. In the middle of the cargo bays were gun jeeps—vehicles
mounted with heavy machine guns—motorcycles, ammunition and
medical supplies. The AH-6 and MH-6 Little Bird Special Operations helicopters of the 160th Special Operations Aviation Regiment
(SOAR)—often referred to as "Task Force 160" or the "Night Stalkers"—were even onboard the aircraft. Relatively small in size, these
special helicopters could be flown up to the rear ramp of the aircraft where they'd land, shut down the engines, have their blades
manually folded back, then be pushed up the ramp on a small set of
wheels.

As the Rangers approached Salinas, the commanders continued

to receive updates, which were not necessarily timely or accurate. A half-hour after departure, Taylor was informed that the runway was definitely blocked. At 11:20 PM, the report was passed that enemy movement had been observed near True Blue. At 12:30 AM, Taylor had learned that the SEALs' insertion on the beach of Point Salines had failed for a second time. By 1:30 AM, the battalion commander made the decision to direct Captain John Abizaid—"The Mad Arab" and future Central Command Commanding General during Operation Iraqi Freedom (OIF)—A Company commander, and his men in the two lead MC-130s to jump.

At 3:30 AM, the reconnaissance C-130 was over the port and the Pathfinder team jumped at 2,000 feet above ground level (AGL). Once on the ground, the Pathfinders confirmed that heavy equipment was blocking the runway. Realizing that A Company's mission required more time than the allocated thirty minutes, Taylor directed at 4 AM that everyone chute-up for a jump.

With only an hour left until H-Hour, the Rangers hastened to perform an in-flight rigging, the most difficult task of an airborne operation when performed within the restricting confines of a combat-loaded transport. More ominously, even though Taylor had received confirmation from the Air Force that the message had been transmitted to the other aircraft in his battalion, the message had not been clearly understood by the last three aircraft, numbers five, six, and seven, of his formation.

In transport number five, Taylor's battalion executive officer, Major Jack Nix, anticipated the jump and directed his men to rig their chutes. Shortly after he had issued the order, though, the Air Force loadmaster erroneously announced the Rangers would be air landing. Amidst much cursing, the chutes were taken off, gathered, and moved forward out of the way.

A little while later, the same loadmaster reappeared, yelling, "Only thirty minutes fuel left. Rangers are fighting. Jump in twenty minutes!" Both the sixth and seventh C-130s had received similar messages and now the chaotic race was on to re-rig for a combat jump. Little time remained and there was no time to follow proper Jump Master Procedures Inspections (JMPI). Employing the "buddy system," teams composed of two Rangers assisted each other to place, adjust and attach the main chute over their battle harnesses, secure the reserve chute, attach large, heavily laden rucksacks to their D rings, and attach weapons cases to the left side of the body. It was a chaotic scene as everyone struggled to be ready. Later studies clearly indicated that the soldier's load was often excessive, with men carrying over 1,000 rounds of ammunition and some rucksacks weighing greater than 120 pounds.

Troubles continued to multiply for the airborne assault elements. With only fifteen minutes to go, the lead MC-130 reported a navigational equipment malfunction that prevented its locating the drop zone in the dark. It also happened that the two lead aircraft were passing through a rain squall, and the lead pilot felt it was not safe to change positions with the tail aircraft.

As a consequence of the MC-130 problems, Major General Richard Scholtes, commander of the Special Operations Task Force 123 (TF 123), who was also in the air in a command EC-130, pushed back H-Hour another thirty minutes to 5:30 AM and directed the two MC-130 aircraft with the A Company Rangers to abort their first run and come in behind the battalion formation. Both actions infuriated Taylor. Not only had he been forced to depart two hours earlier than planned, now he was obliged to jump his battalion ahead of his designated runway-clearing unit. To finish off events, Taylor then learned, as he was attempting to reform the remaining five aircraft that the Rangers on three of them were still struggling to rig for the coming jump. Upon learning of their problem, he directed them to racetrack—fly in a circle overhead—until all of the Rangers were properly rigged.

It was 5:31 AM when Taylor's aircraft began the final approach. Almost daylight, the sky was partly cloudy and the winds were a blustery twenty knots. As the aircraft approached on a straight and level flight at 500 feet and 150 knots, a PRA—People's Revolutionary Army, the name for Grenada's armed forces—searchlight locked onto Taylor's plane from the western end of the runway. Any thought of surprising the enemy had certainly vanished by this time.

From an altitude below the minimum safety altitude of 500 feet AGL—which prompted some Rangers to remove their reserve chutes for there would have been little opportunity to employ it should the main chute fail—Taylor, his headquarters group, and a platoon from B Company exited on the green light at 5:34 AM. Rapidly following each other out the side troop doors of the aircraft, the T-10 chutes opened overhead as the C-130s continued down the drop zone. A 12.7-mm four-barrel antiaircraft machine gun opened up, spewing green tracers across the sky. With the final Ranger exited, the C-130 quickly dove to reduce the enemy's ability to continue tracking and firing on it with heavy anti-aircraft weapons. With engines fully throttled forward for maximum power, the C-130 safely departed the area skimming 100 feet above the sea.

Taylor watched in frustration as the pilots of the next two aircraft—taking heavy machine gun fire—were forced to abort their drops. With only forty-odd men on the ground totally exposed to enemy observation and fire from the surrounding low hills, two hundred yards north of the runway, Taylor called in the two orbiting Spectre gunships for support, thus requiring the troop transports to remain clear of the area.

As communication was being established, the platoon from B Company began to clear the runway of vehicles, obstacles, and debris. Tankers, bulldozer, trucks, and drums were scattered about and stakes had been pounded into the ground with wire strung between them. While some vehicles were found to still have keys in their ignitions, others had to be hot-wired to start. For fifteen minutes, the Rangers cleared the field without a shot being fired in their direction. One operator was able to start a bulldozer and use it to flatten the stakes and push aside the drums.

At 5:52 AM, one of the two A Company MC-130s approached in a hail of tracer fire—motivated it was later unofficially reported by a Ranger officer who had drawn a weapon to threaten the pilots to continue on course—and dropped 115 paratroopers, leaving seven on board unable to get out prior to "red light" and the end of the drop zone. With these additional men on the ground, Taylor ceased clearing operations and moved into a covered position south of Hardy Bay.

A gaggle of C-130s surrounded the southern end of Grenada as the planes jockeyed for position or waited for the 1st Battalion Rangers) to finish chuting up; the pilots remained mindful that they needed enough fuel for the forty-five minute return to Barbados. By 6:34 AM, the remainder of A Company was on the ground, assembling at an alternate site upon discovering a machine gun nest in the vicinity of its original assembly location. The airborne operation continued to drag on until the final 1st Battalion Ranger was on the ground at 7:05 AM.

With the last of Taylor's men on the ground, it was now Hagler's turn. To avoid confusion and to minimize the risk of intermingling the units, he and his men had been forced to wait until all 1st Battalion (Ranger) elements were on the ground. Minutes after the 1st's last pass, the 2nd Battalion (Ranger) arrived overhead in a tight five-plane formation to jump into space at 7:07 AM.

By 7:10 AM, all of Hagler's Rangers were safely on the ground with only two exceptions. One was Specialist 4 Harold Hagen, who broke his leg executing his parachute-landing fall (PLF)—a sequence of movements a paratrooper executes when he hits the ground—on the runway. The second exception was Specialist 4 William Fedak, whose parachute static line became entangled as he exited the door, making him a "towed jumper." Fortunately for him, the Air Force loadmasters were able to safely retrieve him into the aircraft, shaken but not injured. By a stroke of amazing and blessed luck, the American force was on Point Salines without suffering a single casualty inflicted by the enemy.

On the ground, the 1st Battalion (Ranger) moved out to conduct its missions. A Company crossed the tarmac of the runway without a shot being fired and continued to advance on the village of Calliste. During their movement into the village, the Rangers suffered their only killed-in-action (KIA) in securing the runway when an M-60 machine-gunner, Private Mark Okamura Yamane, was struck by a round in the neck after calling out in Spanish for an enemy group who were holed up in a school to surrender. Sergeant Manous Boles then proceeded to lead an attack against the entrenched enemy by driving a captured dozer followed by his squad up the hill toward the school.

A Ranger platoon arrived at the True Blue campus around 7:30 AM. Though there was a brief engagement between the Rangers and a PRA guard detail at the main gate, the campus was secured in fifteen minutes as the PRA soldiers fled north into the hills. This part of the operation proved to be a bona fide success, for not one student had been harmed, none taken hostage, and the Ranger platoon had suffered no casualties.

There was one major problem, though. The Rangers found less than half the number of students expected. The remainder was at a campus called Grand Anse, a site that had not been considered in the operation plan.

On the runway, B Company had been tasked to clear and to secure the control tower and the hills beyond the terminal. For a brief moment, an engagement flared in the vicinity of a Cuban construction camp—Fidel Castro, the dictator of Cuba, had provided arms and approximately forty Cuban military advisors and 635 armed, though mostly older, Cuban construction workers in support of Grenada's totalitarian regime—but those defenders were soon on the run, leaving behind one dead and twenty-two prisoners. Having secured the fuel storage tanks on the high ground 600 yards northeast of the terminal, the Rangers were able to look down into the Cuban mission headquarters at Little Havana and observe two mortars being prepared to fire. Bursts by the Rangers from a captured 12.7-mm machine gun quickly encouraged the Cubans to reconsider their actions and scattered the gun crews. Both A and B Companies of the 1st Battalion (Rangers) worked to consolidate their positions and, by 10 AM, Calliste and the fuel tank hill were secure.

An unfortunate, though fortunately not fatal, incident occurred shortly after 10 AM when two Ranger motorcyclists mistakenly headed for the Cuban mission compound from the terminal. Riding through the valley, the two Rangers came under heavy fire and were both wounded and knocked off their cycles. The members of B Company, 1st Battalion (Ranger), who had observed the scene, were unable to secure the men because of their exposed positions. Eventually, the observing Rangers sealed off the area with sniper fire by mid-afternoon as surrender negotiations were underway and evacuated the wounded men.

Because of the nature of their drop, the 2nd Battalion (Ranger) had been able to assemble quickly and, meeting little resistance, briskly cleared the runway area west of Hardy Bay. A Company moved into the partially completed terminal buildings and up into the old Cuban camp where a small cache of abandoned weapons were found. B Company cleared the narrow strip of land south of the runway towards Point Salines. C Company cleared the low hills from the north down to Canoe Bay. Frustratingly, both Hagler and Scholtes knew that Calivigny would not be secured that day.

The region around Point Salines airfield had been secured. Within four hours of the first drop, the aircraft that had flown in the Rangers were refueled and landing on the airstrip to off load their vehicles and supplies.

OBSERVATION

Operation Urgent Fury is a classic example of events conspiring to threaten the success of a plan. There is an old adage in the military that "The best of plans do not survive the first bullet down-range," and Grenada proved to be no exception.

Urgent Fury, however, was more than an operation that essentially succeeded by default, i.e. essentially executed with hardly any detailed planning at all. Plans are built on intelligence, and there was little of that available—no tactical intelligence to confirm enemy locations, strengths, weaponry, capability, or intents. Plans are created with comprehension of the participant's capabilities but in this instance the two primary service elements, the Army and Air Force, had an exceptionally poor joint-training record that resulted in insufficiencies in C-130 air crews for night-time operations, uncoordinated flight schedules, missing communication equipment, malfunctioning navigation equipment and lastly, erroneous in-flight updates all of which led to mass confusion within the Ranger ranks and aborted drops over the enemy drop zone.

Because of its almost corporate approach to tactics, the U.S. Army fails to instill in its junior leaders that mistakes and unforeseeable factors will lead to chaotic mission execution. Rather than ignoring this fact of combat, leaders should anticipate things going wrong and insure their subordinates are mentally prepared for the anarchy of battle. Only in this way can the practitioners infuse agility into a plan.

The Quran states that "If you don't know where you are going any road will take you there." "Any road" was what the Rangers took that late night and early morning in October 1983. However, Urgent Fury likely being the most poorly planned mission ever executed by the U.S. Army Rangers, the Rangers stumbled on success through remarkable adaptability and sheer drive.

That said, they were also exceptionally lucky for the missing 230 American students and faculty at the Grand Anse campus were not found until the next evening on the 26th (not to mention the 200-plus additional American civilians who were unknown and spread out around the Grand Anse campus). Calvigny, a Day One objective to be taken on the 25th, was not secured until the evening of the 27th. In the end, luck won out, for if the enemy had been better trained, hence more formidable, there would have been a less happy ending. But then again, success in combat is a study in contrasts.

Operation Urgent Fury was the first battalion-level combat mission planned and executed by the Rangers since the end of the Second World War, 38 years prior, and it was nothing to crow about. There was a great deal of rust within the Ranger planning machinery that needed to be scraped off and the planning process itself needed to be reworked. That all did not go well in Grenada proved in the end beneficial, requiring a complete revamp of planning processes that would ultimately revitalize intelligence gathering, training, and coordination between joint services and lead to greater Ranger success in the decades ahead.

'Mass Tac' airborne assault. (Training Picture)

C-141 transport taking off from a secured Point Salines airfield. (Oct. 1983)

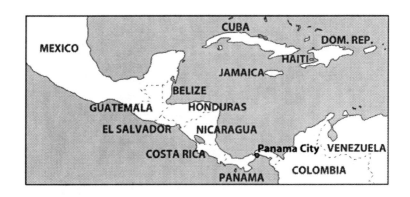

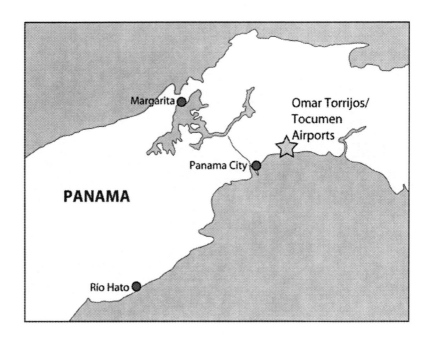

FIFTEEN

¡COJONES!

Victory will come to the side that outlasts the other.
Ferdinand Foch (First World War French Marshall of France
and Allied Supreme Commander, 1851–1929)

DATE: *19–20 December 1989*

LOCATION: *Omar Torrijos International Airport, Panama*

MISSION: *Operation Just Cause - To seize and secure an airfield.*

BACKGROUND

JUST AFTER MIDNIGHT ON 20 December 1989, President George H. Bush authorized the invasion of Panama by U.S. forces in an effort to capture General Manuel Antonio Noriega, the Panamanian dictator, and transport him to the U.S. to face drug-smuggling charges. Among a vast number of other missions to be conducted throughout Panama on the morning of the invasion, OPLAN 90-2 (The standard abbreviation for Operation Plan—OPLAN 90-2 was later named "Operation Just Cause") directed "Task Force Red," the 75th Ranger Regiment, to seize key objectives by airborne assault in the initial phase of the invasion of Panama. Task Force Red was divided into two smaller task forces: Task Force Red-Tango, comprising the 1st Ranger Battalion and C Company of the 3d Ranger Battalion, who were to secure the Omar Torrijos International Airport and Tucumen Military Airfield complex; and Task Force Red-Romeo comprising the rest of the regiment who were to secure the Rio Hato base camp.

By late afternoon of 18 December, the 1st Ranger Battalion was facing weather that had unexpectedly turned exceptionally adverse as a winter storm of rain, sleet, and wet snow gripped the southeastern coast of the United States. The Rangers worked in the open as they went about their duties in 20-degree temperatures—a stark contrast to the 90-plus degrees and 100 percent humidity they would soon be experiencing in Panama. The adversity they faced increased as the weather worsened and the Rangers got wetter, colder and more miserable as they formed along the runway's tarmac at Hunter Army Airfield (HAAF), Georgia, to draw live ammunition, a dead give away that the operation was not just another exercise. (Note: For operational security reasons, only the senior leadership were aware, initially, that this was a "real world" and not a training mission.)

From Hunter, the sixteen aircraft of Task Force Red-Tango departed at 7 PM. Seven C-141s transported the 1st Ranger Battalion while four C-130s transported C Company, 3rd Ranger Battalion, along with the remaining 1st Ranger Battalion elements. Normally for airborne operations, C-130s carry no more than sixty-two paratroopers but, for "Just Cause," these four aircraft were crammed with more than seventy Rangers each. In addition to the eleven troop transports, the air armada included five C-141s carrying vehicles, additional supplies, and equipment.

Completely exhausted from the day's preparation, many of the Rangers fell asleep en route. Two hours out from "green light"—the signal to jump—word was received that the operation had been compromised, and the Panamanians were aware of the Ranger's imminent arrival. For most of the Ranger leaders, they were only surprised that their operation had not been compromised much earlier, given the size and purpose of the overall operation.

At 1 AM on 20 December, after a seven-hour flight, the airborne assault on Torrijos-Tocumen Airport commenced with an AC-130H Spectre and two AH-6 Little Bird helicopters of the 160th Special Operations Aviation Regiment (SOAR) - TF 160 (TF 160) engaging and suppressing the airport's military facilities' defenses with rocket, cannon, and machine gun fire.

Three minutes later at 1:03 AM, according to plan, 732 Rangers of the 1st Ranger Battalion task force exited seven C-141 Starlifters and four C-130 Hercules transports 500 feet above ground level (AGL) over the objective. The seizure of the airport was nearly a flawless operation. By 2:10 AM, twenty-five minutes later than planned as a result of some initial resistance, the 1st Ranger Battalion's objectives on the Tocumen military airfield were cleared of the enemy and secured.

South of the 1st Ranger Battalion, C Company, 3rd Ranger Battalion, was involved in considerably more action. Quickly assembling, the majority of the company had reached its initial objectives within the first fifteen minutes of the assault. The 3rd Platoon had moved to clear the fire station north of the terminal, which was composed of a large rectangular building with two enclosed walkways jutting out of its side running to a rotunda that housed a number of aircraft access gates. South of the terminal, 2nd Platoon was to clear a small Eastern Airlines baggage area, establish a position from where it could provide supporting fires, and then, on order, attack the terminal. 1st Platoon's mission was initially to secure the road entry way into the Torrijos Airport then move to clear everyone out of the terminal's restaurant.

As 3rd Platoon advanced on the fire station, one of the fire trucks moved out from the garage. A squad leader directed warning shots to be fired in front of the vehicle. Deterred by the 5.56-mm tracer rounds cutting the air just feet before him, the vehicle driver did a sharp U-turn and drove the truck back into the garage. Inside the station, the platoon leader found fifteen firemen. Using an interpreter, he convinced them to surrender without a shot being fired.

THE LEGACY

With the fire station secured, the platoon continued on to the main terminal. Shots rang out from the northern rotunda, shattering glass, as the platoon moved across the tarmac. The Rangers scattered under the hail of gunfire. Sergeant Reeves, Specialist Eubanks, and Private First Class William Kelly located some maintenance stairs and entered the terminal.

Inside, the three Rangers observed two Panamanian Defense

Force (PDF) soldiers—who must have fired the shots—run into a women's restroom. The PDF soldiers had started the fight, and the Rangers decided they were going to finish it in what would become one of the strangest five-minutes' worth of close-quarter combat experiences in the annals of Ranger history.

Electing to finish the enemy off with one move, Reeves pulled the pin on a grenade and kicked the restroom door in, only to find a second closed door just inside. With only seconds to spare, he tossed the grenade into the middle of the concourse as he and his men jumped for cover. The detonation blew out what remained of the windows and created a huge hole in the floor.

Gathering themselves, the three-man Ranger assault team led by Reeves proceeded to charge through the two doors, through which only one man could fit at a time. Surprisingly, all was quiet when Reeves burst through the second door into the darkened restroom. Seeing nothing to his right, Reeves was just starting to look towards the stalls on his left when he caught movement out of the periphery of his eye. One of the PDF soldiers was standing on the toilet of the stall closest to the door.

Before Reeves could fire or react, he was struck by three rounds from the enemy's AK-47, fired only three feet away. With two hits to the shoulder, one through the collarbone, and powder burns covering his face, Reeves was knocked to the floor. As he lay on the floor seriously wounded, Reeves was pounced on by the second of the PDF soldiers. Believing he was about to die, Reeves closed his eyes only to be startled and relieved when the enemy soldier and his compatriot quickly disappeared to the rear of the bathroom. Fighting mad and unable to use his right arm to grip his M-16 rifle, the sergeant attempted to grip a grenade with his left but was unable to move his arm enough to get at the grenades in his hip pocket.

Having heard the shots, Eubanks and Kelly crawled on their hands and knees into the dark facility to grab their wounded squad leader. Bullets ricocheted off the walls and floor as one of the PDF jumped out in the open to fire at the Rangers. Three shots bounced off Kelly's Kevlar helmet as they pulled Reeves to safety.

Outside of the restroom, the two Rangers sat the wounded Ranger up against the wall as Eubanks attended to the sergeant's wounds the best he could. Caring little about his wounds and wanting the two PDF soldiers dead, Reeves assisted the two with the development of another plan of attack.

Their plan of action was to start with the toss of another grenade into the room. Opening the second door that had stymied Reeves the first time, the two unwounded Rangers tossed a grenade into the left side of the restroom. Mirrors were shattered and glass flew everywhere with the detonation, but the two enemy soldiers had moved to the far side of the restroom, seeking shelter from the stall partitions located there.

Eubanks and Kelly grew impatient and decided that it was physically impossible for them to get a grenade farther into the room from where they were located. They needed another plan if they were to get these guys and quickly decided that only a personal, face-to-face confrontation would accomplish the mission. Eubanks quietly entered first with Kelly covering the opposite side of the wall. Creeping along the wall as quietly and concealed as possible, Eubanks spotted the two PDF soldiers in the rear of the room. At this point, with no cover for protection and given his increased exposure, the Ranger no longer had the option of tossing another grenade without the possibility of wounding himself.

Raising his weapon and placing one of the enemy in the sights of his Squad Automatic Weapon (SAW), Eubanks pulled the trigger, but the weapon malfunctioned as it failed to chamber a 5.56-mm round. Worse, the machine gun's barrel fell off as a result of the locking lever having become unsecured during their movements. Compromised by the noise, Eubanks had three rounds fired at him from a pistol. The bullets whistled by his head, high and left as he scrambled from the room.

Safely back out in the concourse, Eubanks grabbed and loaded Reeves' M-203 40-mm grenade launcher and secured a second hand grenade. Tossing the hand grenade into the room with the intent to stun the enemy, Eubanks and Kelly rushed through the door and opened fire with their weapons. When they ceased firing, the two Americans incredulously heard the PDF cursing the Rangers and the United States in Spanish.

Understanding and able to speak a little Spanish, Eubanks told the two PDF soldiers to lay down their arms and surrender. With each offer to surrender, one of the enemy soldiers would poke his head around the far corner and yell at Eubanks, ("Fuck off!"). On the third cursed challenge, one of the PDFs poked his head out a little too far, and Eubanks was able to fire a single round through the heckler's neck. The wounded Panamanian dropped his weapon and crumbled to his knees. Eubanks screamed at him to lie face down but the babbling and dazed PDF soldier was not listening. M-203 in hand, Eubanks grabbed the wounded PDF soldier by the back of his shirt and pushed him to the floor.

Neither Eubanks nor Kelly saw the second PDF soldier behind a stall door. Lunging for Eubanks' weapon, the second enemy soldier struggled for the grenade launcher as the wounded soldier on the floor rolled over and attempted to pull a pistol from his waistband. Eubanks and Kelly were able to kick the wounded soldier out a window where he bounced off a ledge onto the tarmac twenty-five feet below.

Having somehow survived the plunge, the PDF soldier's luck finally had run out, for he had fallen in front of a Ranger M-60 machine gun position. Refusing to halt as ordered—the only viable act that would have saved his life, he was finally killed with a burst of 7.62-mm rounds.

Inside the restroom, however, the struggle still ensued. Able to get both hands on Eubanks' weapon, the other wounded enemy soldier attempted to wrestle the weapon away from the Ranger rather than shoot him with it. Enraged, Eubanks pushed the Panamanian against a urinal and began to kick him repeatedly, screaming for Kelly to shoot the man. Kelly's shot to the arm was immediately followed by two more to the head that finally brought the action within the airport restroom to a close.

For their actions, Eubanks and Kelly were awarded the Bronze Star with "V" device (for Valor) while Reeves was awarded the Army Commendation Medal (ARCOM) for his supporting contribution.

OBSERVATION

In the heat of battle, in the fog of war, life-altering events occur that could end differently with a tad more communication or thought. For example, trapped as they were, the two PDF soldiers certainly had the option to surrender but they may not have understood Eubanks' limited Spanish. On the other hand, they just may have easily been fools who believed they could fight their way out of what they had to know was an American invasion.

Given the circumstances, there are those who may question the tenacity and perseverance of this small Ranger group when the situation, if it had been given just a bit more thought, may have dictated a wait and see option that could have resulted in the PDF soldier's eventual surrender. Often the exigencies of the moment permit no cool calculation but demand reactions in a series of expedients. And, when that occurs, he who is better trained is generally the eventual victor.

In combat, one generally seeks and maintains superiority by engaging and destroying the enemy at the greatest range possible. This enhances one's own protection. Closing with the enemy to the point of hand-to-hand combat when there are still long-range options—and time—available negates much, if not all, of that superiority.

Furthermore, should one decide that close-quarters combat is the only alternative, be certain that your weapon is properly functioning. Eubanks' SAW not only failed to chamber a round correctly when he had an opportunity to finish the task, but this happenstance also demonstrated the more embarrassing gaff of the weapon falling apart at the same time. Equipment failure as a result of a lack of maintenance is inexcusable. It can be the difference between life and death or mission accomplishment.

In the end, if there is one lesson to be derived from this engagement, it is that patience above all, can be a Ranger virtue. And, while some could decide that this engagement, after the fact, was foolish and risky, no one would dispute that these Rangers certainly had ¡Cojones! . . . or balls!

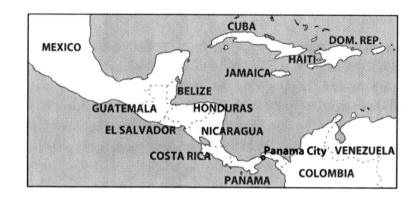

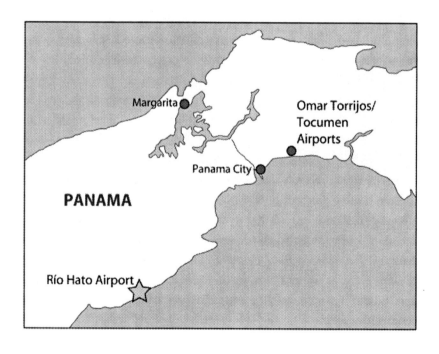

SIXTEEN

CONFIDENCE

Fields are won by those who believe in winning.
— **Thomas Wentworth Higginson**
Union Colonel American Civil War and writer, 1823–1911

DATE: *19–20 December 1989*

LOCATION: *Rio Hato Airfield, Panama*

MISSION: *Operation Just Cause - Airfield seizure and destruction of local enemy forces.*

BACKGROUND

TASK FORCE RED-ROMEO'S OBJECTIVE was Rio Hato, home base to the 6th and 7th Panamanian Defense Force (PDF) Rifle Companies—considered the best fighting units in the PDF—and located approximately sixty miles southwest of Panama City on the coast of the Gulf of Panama. Intelligence estimates placed the combined two-company strength at 440 soldiers with an additional fifty-man engineer platoon. It was also believed that there were up to 250 cadets at the Panamanian noncommissioned officers (NCO) Herrera-Ruiz Academy. The facilities at the airfield included two company and one platoon barracks, a communications center, a training center, two motor pools, an ammunition supply point (ASP), a medical dispensary, and the operations complex for the 4,380-foot long runway. General Noriega's beach house, Farallon, was also located nearby.

Estimates indicated a wide variety of weapons capability. For mobility, there were three V-300 and sixteen V-150 armored cars and fifteen motorcycles. Three ZPU-4 14.5-mm quad (four) barrel air defense guns and at least forty-two machine guns could provide anti-aircraft fires. Heavy direct fires existed in the form of nine bazookas, four recoilless rifles, and up to 200 RPG-7 rocket-propelled grenades. Indirect fires could be provided by at least twenty-three mortars. Overall, the mission facing the Rangers was not an easy one.

Designated "Area of Operations Eagle," or AO Eagle, Rio Hato was divided into two different operational zones for each attacking Ranger battalion, each region having two objectives within it. The southern region was assigned to the 2nd Ranger Battalion and included the 6th and 7th Company compounds—named for identification purposes as Objectives "Cat" and "Lion." The two companies of the 3rd Ranger Battalion were to secure the northern region of Eagle to include the NCO academy, the camp headquarters, the airfield operations complex, the motor pools, the communications center—all included within Objectives "Dog" and "Steel"—and the ammunition supply point at the far northern end of the runway. Additionally, the battalion was to clear the runway for air landing operations to follow and cut the Pan American Highway that ran through the area.

The weather at Lawson Army Airfield, Georgia, turned bitterly cold. Housed in a tent city, the Rangers assembled in the Airborne School hangars just hours prior to their departure, where they began the demanding task of rigging their chutes, conducting Jump Master Pre-Inspections (JMPI), and organizing the "chalks"—having the soldiers line up in the order they would load the aircraft and jump. They were bathed in red light to prepare their eyes for nighttime operations and to minimize their exposure to the PDF by having white light shine outside the aircraft through open paratrooper doors as they flew towards and over the drop zone.

Rangers walked down ample rows of opened ammunition crates, drawing live ammunition, hand grenades, and other munitions with which they would jump into combat. Morphine, for pain control and a controlled substance, was distributed only to the medics. Chaplains moved about the airfield conducting voluntary services for the Rangers. Those who observed the process would note later that the Rangers "did not seem frightened or anxious, just very aware of what was about to happen, and confident of their abilities."

The weather continued to worsen as freezing rain continued to fall. Exposed to the elements, the Rangers "dined" on chicken soup and cookies and were issued heavy wool blankets to ward off the chill. Prior to loading and departure, the regimental commander, Colonel William F. Kernan, spoke to his assembled Rangers standing in mud and slush. Reminding them of the rehearsals they had so successfully and confidently completed just days prior, the senior Ranger promised them this: "We're going down there with everybody here and we're coming back with everybody here."

Tired, soaked, chilled, and slightly apprehensive, the Rangers entered the aircraft "loose-rigged" in their chutes and packed themselves into the heat and humidity of one of the thirteen passenger C-130s. The flight, which included two heavy-drop C-130s—palletized vehicles, heavy equipment and supplies, departed soon thereafter. During the long flight, a five-gallon water can served as a "chamber pot" and was passed up and down the heavily crowded aisles of the transports.

As the Ranger task force neared Panama, the Rangers elbowed each other awake and prepared to stand up and hook up their parachute static lines to the anchor line cables at six minutes out from the drop zone. Following the final jump command of "Sound off for equipment check"—when each Ranger in the chalk taps the man in front to confirm his equipment is safe to jump, each planeload of Rangers proceeded to recite the memorized and honored lines of the Ranger Creed.

THE LEGACY

The attack on Rio Hato commenced simultaneously with the assault on Omar Torrijos International Airport and Tocumen Military Airfield complex by the 1st Ranger Battalion and C Company of the 3rd Ranger Battalion—designated as Task Force Red-Tango. But instead of opening up with overwhelming firepower against the barracks,

two F-117A Stealth fighters flown from Nellis Air Force Base at To-
nopah, Nevada, flew over the airfield at 1 AM. Each carried a 2,000-
pound bomb with a time-delay fuse that was to be dropped on the
7th Company headquarters and next to the 6th Company headquarters. Lieutenant General Carl Stiner, commander of the XVIII Airborne Corps and war-fighting commander of Joint Task Force South
to which the Rangers were attached, had made that decision to minimize collateral damage and not destroy the base or kill large numbers of Panamanian soldiers. Instead, he opted to stun the enemy and
create a period of confusion by having the Air Force deliver the two
one-ton bombs near the command and control centers.

The method of delivery was determined by the Air Force which
selected the newest aircraft in its inventory. Embarrassingly, the two
aircraft were wide of the mark on their first wartime deliveries. The
lead Stealth, approaching the target without detailed targeting information, dropped its payload 160 yards northwest of the intended
point of detonation. The second Stealth, following the first's lead,
dropped its bomb between the two company buildings. Despite the
noise and destruction, the bombing strikes accomplished nothing for
the airborne assault had already been compromised hours earlier.
The buildings were empty.

A contingent composed of an AC-130H Spectre gunship, two AH-
64 Apache attack helicopters, and two AH-6 Little Bird helicopters
belonging to TF 160, opened fire on pre-selected targets immediately
upon detonation of the second bomb. For ninety seconds, machine
gun, cannon, rockets, and Hellfires lit up the sky as they impacted on
a wide variety of targets, especially the ZPU-4 air defense guns.

The airborne force approached north to south from over the Gulf
of Panama to be greeted by tracer rounds fired skyward. Despite
the fires, three minutes after the initiation of the Stealth's attack, the
green light of the lead C-130 came on and the first of 837 Rangers
exited the blackened aircraft at 500 feet above ground level, though
not all Rangers were able to exit on the first pass with the drop zone
(DZ) so tight.

The anti-aircraft fires were heavy and momentarily unanswered
as the Spectre and helicopters had to cease-fire and withdraw while

the paratroopers were in the air. Eleven of the thirteen "pax" (passenger) transports were hit as they over flew the objective. The two heavy-drop aircraft followed close behind to deliver their eight pallets loaded with jeeps, motorcycles, and supplies.

Inside one C-130, a 7.62-mm bullet penetrated the skin of the aircraft and struck Staff Sergeant Rich Wehling of B Company, 3rd Ranger Battalion, above his flak jacket and in the back, as he was shuffling to the door. As he staggered forward and fell to the floor, the Ranger behind him cut his static line, pushed him aside, and continued to lead the remainder of the chalk out the aircraft door into sweltering Panamanian night air.

Wehling, after regaining consciousness, found himself all tangled up in deployment bags. The aircraft's pilot, Colonel McJunkins, made his way to the rear of the plane to apply first aid.

When asked by the colonel, "How do you feel?" the Ranger's response was "Mad, sir."

Surprised by the reply, the pilot asked why, to which Wehling remarked, "I didn't get to jump with my buddies."

Back at Rio Hato, Wehling's fellow Rangers found themselves in a 360-degree firefight on the ground. The 2nd Ranger Battalion's section of the drop zone was rough terrain and at least a few of the Rangers found themselves hung up in trees. Metal pickets (special pipes on which to hang wire) with barbed wire were located in the northern sector and there were energized power lines running down the middle of the area. One Ranger who was unable to avoid the power lines was Kernan, the regimental commander. As he approached the ground, he passed through some power lines and became hung up on a fence. His chute, tangled in the overhead wires, caught on fire. Suspended ten feet above the ground, the colonel released his harness and climbed down as his burning parachute shorted out the wires and cut off all electricity to the camp. As he would later comment, he had the "lightest landing at Rio Hato."

One Ranger, Private First Class John Price of the 2nd Ranger Battalion fell to his death when the chute malfunctioned. Thirty-five other Rangers—almost five percent of the force and considered acceptable losses in an airborne operation—were injured on the jump, including the 2nd Ranger Battalion commander who suffered a broken foot. Another Ranger officer, Captain Steven G. Fogarty, suffered two broken ankles. Staff Sergeant Richard J. Hoerner landed in the middle of a road. As he struggled to get his parachute harness off, Hoerner observed an enemy vehicle rapidly approaching his location. Unable to get his M-16 rifle out of its case in time, the vehicle sped by him, catching his partially inflated chute. Slammed to his back, the Ranger was dragged for nearly one hundred feet before he could pop the cable loop assemblies of his risers (the device by which a paratrooper can disconnect his parachute from the harness that is worn on the body). Back on his feet, the Ranger set about gathering up essential equipment that had fallen off as he'd been drug before making his way back to his fellow Rangers.

On the ground, the Rangers found an alert and dispersed PDF which, while not well organized, were all over the place, firing at the Rangers as they discarded their chutes. As the Rangers moved to assemble, two things kept them from firing on each other in the dark. One was the word, "Bulldog," which was used as a sign/countersign. The other was the distinctive ragtop camouflage of their Kevlar helmets—strips of camouflage material attached to the helmet to breakup its distinctive shape, which the Rangers wore to differentiate themselves from some of the PDF soldiers who also wore American-style helmets.

The Rangers of A and B Companies, 3rd Ranger Battalion, moved to their objectives. A Ranger group composed of Lieutenant Loren Ramos, Staff Sergeant Wayne Newberry, and Specialist Oler expected to find a 25-foot-high stone structure at the main gate, but found instead a two-story, arched-structure with several rooms at ground level and castle-like turrets on top. Worse still there was a .50-caliber machine gun somewhere on the structure firing on them as they moved toward it.

They made it safely to the gate. But, just as two of the Americans had stepped into the road to assault one of the rooms, a PDF V-150 armored car sped by them forcing them to dive for cover. As the vehicle vanished along the Pan American Highway, the Rangers turned their focus back to the job at hand, eliminating the machine gun, killing two of its crew, and clearing the gate structure.

As other Rangers fanned out to the northwest toward the ammunition supply point, Staff Sergeant Olivera jumped into a ditch to avoid some incoming small arms fire. Unfortunately for him, two PDF soldiers were in the ditch, one of whom immediately shot and seriously wounded the Ranger in the shoulder and chest. Literally cheering over their success, the two enemy soldiers proceeded to cut off the American's patches as war trophies, including the Ranger scroll on his left shoulder. Obviously hoping to leave him dead, one of the Panamanians fired a pistol pointblank at his head. Prior to departing, they then wrapped a black bandanna that read Machos del Monte, "Men of the Mountain," around his rifle and departed the area.

Fortunately for Olivera, the executioner's bullet had been deflected enough by the sergeant's Kevlar helmet to only ricochet off his skull and exit from behind the Ranger's left ear. He regained consciousness around 6 AM. Unable to move either his arms or legs initially because of the large loss of blood, the sergeant eventually maneuvered his body enough to contact a rescue team via a portable radio he grabbed from his rucksack. This act most likely saved his life.

In the southern portion of the airfield, A and B Companies of 2nd Ranger Battalion were flushing out the enemy, moving from building to building, conducting room-to-room sweeps as they did. Company A worked its way through the Non Commissioned Officer (NCO) Institute while B Company, entrusted with the main effort, cleared the complexes of the PDF 6th and 7th Rifle Companies, a mission it had spent considerable time training for back in the States. As the Rangers pressed forward, the PDF would abandon the building but hide in the cluttered terrain and ambush the Rangers as they advanced to the next building. Specialist Philip Lear was killed in the attack on the barracks.

The Rangers did not fire or toss grenades indiscriminately. During the sweep of one barracks building, a Ranger recognized that there were unarmed PDF within. Taking a chance, he entered a room without prepping it only to find 167 cadets, ages fourteen to eighteen, cowering inside. Rounded up and searched, the cadets sat out the remainder of the fight.

C Company, held in reserve, dispatched a small force of Rangers before dawn to secure the Farallon beach house. Encountering a detachment of bodyguards that fired on the Rangers from the rooftop, the Rangers opted not to waste much time or effort in trying to coax the enemy out. Firing a Light Anti-tank Weapon (LAW)—an explosive rocket, they proceeded to blow away the front door. The guards fled but were captured later.

On the ground, Kernan and his regimental headquarters, "Team Black," were located in a recreation center on the north end of the airfield from where they coordinated the various supporting fires and served as the task force's liaison with Stiner. Seven minutes ahead of schedule, one hour and fifty-three minutes into the operation, Kernan reported the airfield secure, and authorized the first aircraft landing of soldiers and supplies.

The fight had cost the Rangers four KIA. Twenty-seven other Rangers were wounded in battle. In exchange, the PDF suffered thirty-four dead, an unknown number wounded, and 362 taken prisoner along with forty-three detained civilians.

OBSERVATION

Operation Urgent Fury, the invasion of Grenada, was a demonstration of how not to plan and execute an airborne invasion. Operation Just Cause, on the other hand, was a demonstration of good planning and trained execution. Invaluable lessons had been learned in

October 1983 that served the Rangers well in December 1989. Though the Panamanian operation was much greater in scope and covered a larger geographical footprint against a better armed and prepared enemy force, the overall Ranger experience in Panama was relatively mild in contrast. Though that experience was not particularly thrilling or exciting, per se, it was the product of the Rangers confidently executing an airborne assault that had been adequately planned, and that they had been properly trained to execute. Such confidence in one's plan and overall abilities usually leads to success.

In contrast to Grenada, the United States had been planning for the possible invasion of Panama for over a year. Units from the 7th Infantry Division (Light) and the 82nd Airborne had used the year to rotate through jungle school and conduct reconnaissance during the course when time allowed. Topographical maps and information booklets on the PDA, including key Spanish phrases, were assembled well in advance. The collapse of the PDF can be attributed to the swift and assured assaults of the U.S. task forces. Clearly stunned by the rapidity and professional execution of the invasion, PDF resistance vanished within 24 hours.

While tightly packed within the confining constraints of a tactical transport, Rangers apply the green and loam face paint from a "camo stick" in preparation for their airborne assault in Panama. (19–20 Dec. 1989)

Members of the PDF—Panamanian Defense Force—are secured and kept under guard. (20 Dec. 1989)

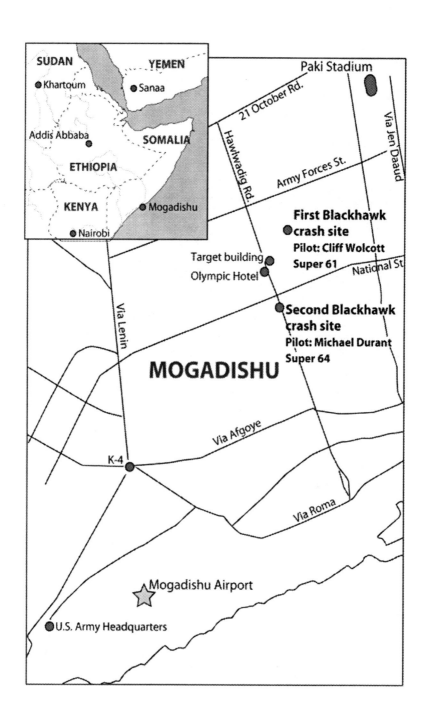

SEVENTEEN

SACRIFICE

The choice of death or dishonor is one which has always faced the professional fighting man, and there must be no doubt in his mind what his answer must be. He chooses death for himself so that his country [or comrades] may survive, or on a grander scale so that the principles for which he is fighting may survive.

—Sir John Cowley
British Lieutenant General, 1905–1993

DATE: *3–4 October 1993*

LOCATION: *Mogadishu, Somalia*

MISSION: *United Nations Operations Somalia II (UNOSOM II). Capture two senior lieutenants, Muhammed Hassan Awale and Omar Salad Elmi, of the the main Somali Warlord, Mohamed Farrah Aidid, deep within a city embroiled in civil war.*

BACKGROUND

FOR YEARS, THE WORLD looked on as a bitter civil war among six rival clans was destroying Somalia and brought about a famine that was killing tens of thousands of innocent people a month. The once beautiful capital and port city of Mogadishu lay in ruins from five years of civil war. Under the directive of the clans, roving bands of technicals—teams of Somali fighters in small trucks with mounted heavy weaponry—controlled all movement of goods into and throughout the country. Clan chiefs—warlords—hoarded food and medical supplies, dispensing them only to their supporters, thus denying basic sustenance to the vast majority of the population.

Death and misery escalated. CNN caught it all on tape, transmitting it around the world. Drought exacerbated the effects of famine. By the summer of 1992, over 300,000 Somalis had perished during the preceding eighteen months. The United Nation's efforts at simple famine relief failed miserably as the technicals stole most of the earmarked supplies. The U.N. Secretary General, Boutros Boutros-Ghali, appealed for military assistance from member states and initiated the United Nations Operation in Somalia (UNOSOM) effort. But only a token force was raised and lack of U.S. involvement ensured failure.

On 16 August 1992, President Bush authorized Operation Provide Relief, a token airlift with a small support base. By November, however, Operation Provide Relief wasn't any more successful than previous U.N. efforts; civil war and famine continued to take their toll. Now, nearly a half-million Somalis were dead, up to a million more would die from hunger if the status quo were maintained, and almost one million had fled the country. It was feared that the Somali struggle might spread throughout Africa like cancer.

In late 1992, following his defeat for reelection, President George Bush initiated direct military intervention in the region. The Chairman of the Joint Chiefs of Staffs, General Colin Powell's, employed his "combat model": while the mission would be limited to humanitarian relief for the Somalis, overwhelming force would be used to crush or frighten off any armed resistance. The orders were issued to U.S. forces on 25 November. On 3 December, the U.N. Security Council passed Resolution 794, authorizing under Chapter 7 of the U.N. Charter, the use of U.N. force to defeat a threat to international security, "to restore peace, stability, and law and order with a view to facilitating the process of a political settlement under the auspices of the United States." A great void opened between what the United States intended and what the United Nations directed, for while the U.S. was committed to a limited role where one inhibits fighting, the U.N Resolution had the world body going after bigger game—peacemaking, a much greater and more dangerous task.

In the early morning hours of 9 December, elements of the Marine Fleet Recon and Navy SEALs, anticipating the potential for resistance, attempted to make their way ashore secretively and tactically at night near the port of Mogadishu, only to be met by a disquieting mob of journalists, cameramen, and bright lights. Referred to as "Operations Other Than War" (OOTW), the force had sufficient strength to bring organization and relief to the people of Somalia as part of United Nations Operation Somalia (UNOSOM) II. A total of 25,426 American soldiers, seamen, and airmen—designated Unified Task Force (UNITAF)—were assigned to Operation Restore Hope with nearly 16,000 serving on the ground in Somalia.

UNITAF began to leave Somalia on 4 May 1993, the Bush-initiated operation having been successful. President William J. Clinton was now in the Executive Mansion. But many knew that the seeming order in Somalia was only an illusion. The warlords had lain low, waiting for the American-led U.N. coalition to depart. Once gone, the clans would rule again. The American/U.N. operation had only been an interlude with no final solution. "Peacemaking" had failed.

With the departure of UNITAF, the United Nations' effort under the command of Turkish Lieutenant General Cevik Bir began to work on the nation-building effort with a mix of 28,000 U.N. soldiers, which included 4,000 Americans. Unfortunately, Somalian warlords saw these humanitarian nation-building actions—and the combat forces associated with them—as a threat to their power base, and the situation quickly deteriorated as the warring tribes shifted their aggressive attentions from each other to the United Nations forces. The main Somali Warlord, Mohamed Farrah Aidid—"Aidid" in Arabic means "one who tolerates no insult"—was one of the warring chieftains and head of the Habr Gidr clan. His militia, the Somali National Alliance (SNA), was more disciplined than most. In an interview, Aidid blamed the U.N. forces for interfering in the internal affairs of his country and for blocking his clan's succession to power. "We intend to rule" was his edict.

Anti-U.N and anti-American rhetoric gave way to full-blown violence by 5 June when Aidid, forewarned by traitorous sources in U.N. headquarters, ambushed a careless and lightly armed Pakistani force preparing to seize his pirate radio station. The ambush killed twenty-four and wounded fifty-three, including three American wounded in action (WIA) from the QRF—Quick Reaction Force—which deployed to cover the Pakistani withdrawal. On 6 June, the U.N. Security Council—with U.S. sponsorship and approval—passed Resolution 837 calling for the apprehension "for prosecution, trial, and punishment" of those responsible for the ambush and to use "all necessary measures" to install United Nations authority "throughout Somalia." It was determined that Aidid and his SNA were responsible for the ambush, and a plan was developed to capture him.

U.N. intelligence estimated that Aidid's SNA militia was composed of 1,000 regulars. Some estimates placed the figure at 6,000, though the actual number may have been as high as 12,000. Most important was the fact that whatever the final total, the militia was a formidable force with the advantage of operating on familiar ground and the potential swift augmentation of "irregular" volunteers. Nearly every male in Somalia owned his own weapon, which was predominately the 7.62-mm Soviet-made AK-47 assault rifle. For these men, and many others, strife and fighting were the norm, and the traditional warrior gained high nobility within Somali society.

U.N. ambassador to Somalia, U.S. Admiral (Retired) Jonathan Howe, eventually requested 1st Special Operational Detachment-Delta—the premier three-squadron U.S. counter terrorism unit (popularly known as "Delta Force")—to assist in Aidid's capture. Though there were American soldiers of the QRF in support, they, as with most purely combat elements, were not skilled in the "snatch-and-grab" tactics necessary to capture Aidid.

Aidid's "war" against the United States continued, resulting in the eventual deployment of Task Force Ranger, a 450-man force composed of approximately sixty men from the 150 men assigned to Squadron C, 1st Special Forces Operational Detachment-Delta located at Fort Bragg, North Carolina; B Company (Reinforced), 3/75th Ranger Infantry Battalion from Fort Benning, Georgia; and support helicopters from the Army's 1st Battalion, 160th Special Operations Aviation Regiment (SOAR)—the world's finest night fliers known as the "Night Stalkers," stationed at Fort Campbell, Kentucky. Commanding Task Force Ranger was Major General William F. Garrison, a combat veteran and former commander of Delta.

Frustration quickly set in for the task force after its arrival on 26 August as its target had last been spotted nearly a month earlier on 28 July. The problem facing American planning was that Mogadishu society was a perplexing, convoluted web of interlocking family and kin, a base of support into which Aidid could vanish for days, months, or even years. Making matters worse were the Somali agents that served as human intelligence (HUMINT) sources in the city. Though the Joint Special Operations Command (JSOC) controlled the Somalia agents through the Intelligence Support Activity (ISA), the agents were fearful of venturing into Mogadishu's menacing streets after nightfall—and night operations were the task force's specialty and a strength. Despite these specialties and strengths, however, the task force failed to capture Adid from 30 August to late September.

THE LEGACY

The seventh and final mission of Task Force Ranger commenced at approximately 1 PM on 3 October when a Somali agent passed word that a number of Aidid's lieutenants, including two of the six on the expanded target list—Muhammed Hassan Awale and Omar Salad Elmi—would be meeting later that afternoon. A Little Bird helicopter had been dispatched to observe the agent from a distance through a telescopic lens. Using a pre-designated signal, the agent stopped his vehicle in front of a house, raised and lowered his hood, and then drove away. The aerial observer marked the spot on his street map.

The target was in the vicinity of the Olympic Hotel—a white, five-story building that served as a landmark since it was one of the few large buildings left intact in the city. Hawlwadig Road, intersected by narrow dirt alleys, ran in front of the hotel and was one of the few paved roads in the city. Across Hawlwadig, one block north, was—as it turned out—the ultimate target house, a two-sectioned building with two stories in the front, three stories in the rear, and a flat roof on both. L-shaped, the structure had a small courtyard enclosed by a high stonewall.

Just three blocks to the west of the hotel was the Bakara Market, the most heavily armed region of Mogadishu. This area was known by soldiers as "the Black Sea" and was referred to as real "Indian country" by the Ranger Task Force commander, Major General Garrison, when briefed of the target location. Earlier, he had stated, "If we go into the vicinity of the Bakara Market, there's no question we'll win the gunfight, but we might lose the war." Garrison envisioned what would transpire if events did not go smoothly and the task force became bogged down deep in the city. Surrounded deep in enemy territory, the Americans would need overwhelming firepower to keep the Somali militia at bay and to clear a path for extraction to safety. Such massed firepower could only result in immense death and destruction within the confines of an urban area. The general recognized that this state of affairs would undermine the campaign for the hearts and minds of the Somalis.

Despite the known risks, the operation commenced using the code word "Irene." President Clinton had been briefed on each of the previous missions in advance. This one would be different, though, for it was a "target of opportunity" . . . one that just happened, not planned . . . but required swift execution. Thus, neither the president nor his staff was aware of the ongoing operation at the time of execution.

The assault force was formidable, consisting of seventy-five Rangers and forty Delta soldiers inboard an air armada of sixteen helicopters. Wearing desert camouflage battle dress uniforms (DCU) and flak vests, the Rangers had elected to travel lightly, leaving behind bayonets, canteens, and night-vision devices (NVD)—believing this was to be a rapid daylight operation. Communications would be transmitted in the clear, meaning they would not be encrypted and could thus be monitored by the Somalis with the proper equipment—meaning military surplus or off-the-shelf commercial scanners.

The Delta and Ranger assault force would be inserted . . . dropped into place . . . by four MH-6 (Military Helicopter) and six MH-60 Black Hawks with four AH-6J (Attack Helicopter) Little Birds providing close air support. Tiny, quick, and highly maneuverable, the bubble-front helicopters were an exceptional asset. The AH-6Js were armed with mini-guns and a Hydra 70-mm rocket system. The MH-6s carried benches mounted on the outside skids on which Delta operatives rode.

The Black Hawks employed by the 160th SOAR Night Stalkers (also referred to as "Task Force 160"—TF 160) were MH-60A and MH-60L models. The former were UH-60s (Utility Helicopter) with Forward Looking Infrared Radar (FLIR) installed; the latter had not only the FLIR but upgraded engines and weather radar as well. For ease of movement and greater visibility, the pilot and copilot's doors were removed. The SOAR also operated "K" models that had an aerial refueling boom, long-range tanks, and satellite communications capabilities, in addition to the other items found on the "L" models. All of the TF 160 Black Hawks were configured to burn any type of aviation fuel.

A combat search-and-rescue (CSAR) MH-60 with a special team inboard composed of Rangers and Air Force Para-Rescue Jumpers (PJs)—in essence a flying reserve that was capable of more than just routine search and recovery—stood by while the air mission commander, Lieutenant Colonel Tom Matthews, the SOAR commander, and Lieutenant Colonel Gary Harrell, Squadron C commander, callsign Romeo 64, circled the operation 3,000 feet overhead in a command and control (C2) MH-60. A Navy P-3 Orion aircraft—with a Magnetic Abnormality Detector (MAD) inboard that typically was used to hunt submarines rather than large weapons caches—and a MH-60K were also on station to videotape and photograph the action for post conflict—after action review and intelligence gathering purposes.

Since the target area was too restrictive and too dangerous to land helicopters for the extraction of prisoners—"precious packages"—an assault force, comprising a fifty-two-man Ranger ground element as well as some Delta operatives and Navy SEALs from SEAL

Team Six supported the operation from its base at the airport, three miles from the target. The 3rd Ranger Battalion Commander, Lieutenant Colonel Danny McKnight, call sign Uniform 64, controlled the entire operation. The twelve-vehicle convoy was composed of seven Kevlar-lined and ballistic-armored Humvees, mounted with MK-19 grenade launchers or .50-cal machine guns, two unarmored cargo Humvees, and three five-ton cargo trucks. All of the vehicles were lined with sandbags for additional protection from small arm fires and fragmentation. In addition, the cargo trucks had additional improvised protection with dirt-filled wooden ammunition crates installed along the flatbed sides.

The helicopters lifted off at 3:32 PM, a thirty-seven minute delay after information revealed the target house was actually a block farther west of the original location; the agent having been too fearful of stopping directly in front of the objective. Taking a circuitous flight from their staging base just three miles away from the objective and moving low and fast over the ocean's breakers, the aircraft made a dash over the city, with the MH-6s carrying four Deltas, two to a side, on their external benches. With the aid of aerial surveillance, Garrison and his staff in the Joint Operations Center (JOC) command post, located in a dilapidated building at the Mogadishu Airport, followed the operation as two AH-6's flying advance guard security buzzed the target house ahead of the approaching main force at 3:43 PM.

The AH-6 sweep revealed nothing abnormal. Following close behind, four MH-6 Little Birds and two Black Hawks inserted forty Delta soldiers under the command of Captain Scott Miller on the Hawlwadig road amidst a billowing swirl of rust-orange dust in front of the building.

The MH-6s carrying their external troops made their runs first. The lead aircraft, Helo 1 code named Star 41, determined that it was unable to land in the tightly constricted area and "fast-roped" its Delta operatives to the ground from a rope attached to the helicopter. The pilot of Helo 2 realizing that his landing zone (LZ) was also too small, landed on the far side of an intersection.

The combined downdrafts of Helo 1, which had landed in an intersection following the disembarkation of its load and Helo 2 had created such a brownout that Helo 3 was forced to do a "racetrack" . . . circling overhead and returning for another try. Helo 4, the trail bird, did not hear Helo 3's transmission that it was going around. It continued to the LZ, using the building to the left and some wires directly across the front as reference points.

Adding its own dust to that which already existed, Helo 4 quickly found itself in a total brownout. From its hover, the helicopter set straight down. Disoriented by the brownout and noise, and thinking that it was still in a hover, the left side team lowered its rope and conducted the world's shortest fast-rope on record given that the aircraft was already on the ground. As Helo 4 took off and exited the area through the billowing clouds of dust and dirt, the crew was surprised to see that it had flown directly over Helo 3, which had landed just a few feet away.

Following the four Little Birds, the two Black Hawks fast-roped their passengers of twelve Deltas each. Once safely offloaded, the teams stormed the target building with team leader, Sergeant First Class Matthew Rierson, coordinating the roundup.

Simultaneously, sixty Rangers under the command of Captain Mike Steele, call sign Juliet 64, company commander of B Company, 3rd Ranger Battalion, were to be inserted into the objective to establish a security perimeter encompassing the four corners of the target's city block. The four Black Hawks, each armed with two six-barreled 7.62-mm mini-guns capable of firing 4,000 rounds per minute from each door and carrying fifteen Rangers each, approached their designated insertion points. The Rangers aboard were armed with 5.56-mm M-16A2 assault rifles, 5.56-mm M-249 Squad Automatic Weapons (SAW), 40-mm M-203 grenade launchers—attached under the barrel of the M-16, the 7.62-mm M-60 machine gun—affectionately, or derogatorily, referred to as the "Pig" depending on one's perspective and experience with the weapon, M-72 Light antitank Weapons (LAW), and a various assortment of grenades. The 5.56-mm ammunition were green-tipped, armor-piercing rounds with tungsten carbide penetrator cores—excellent for punching holes in metal. Unfortunately, armor-piercing rounds passed through flesh without fragmenting and hence did not expand like a mushroom thus tearing out large pieces of muscle and flesh. In combat terms, these 5.56-mm rounds lacked "stopping" power.

Chalk One—the term "chalk" being the designation for a group of passengers transported on one aircraft—under the command of First Lieutenant Larry Perino, 3rd platoon leader of B Company's 1st Platoon, and carrying the company commander, Captain Steele, fast-roped into a street intersection one block west on Hawlwadig Road. Chalk Two fast-roped into the northeast corner under the command of First Lieutenant Tom DiTomasso, B Company's 2nd Platoon leader. Chalk Three, under the command of Sergeant First Class Sean Watson and part of Perino's platoon, fast-roped into the southwest corner on Hawlwadig Road.

To the northwest, Super 67 with Staff Sergeant Matt Eversmann's Chalk Four (part of DiTomasso's platoon) fast-roped a block north of its proper position on Hawlwadig Road. With an inauspicious beginning, the mission's first casualty was suffered when eighteen-year-old Ranger Private Todd Blackburn ended up unconscious, bleeding from his nose, ears, and mouth on the dirt road beneath the whirling blades of the Black Hawk hovering above. He had jumped but failed to secure a firm grasp of the rope, resulting in at least a forty-foot plummet to the ground. The chalk's medic, Private Second Class Mark Good, immediately began to administer first aid to the seriously injured Ranger.

Meanwhile, the Delta team operatives had seized and handcuffed the mission's "precious cargo," twenty-four prisoners, including the two primary targets. At the twenty-minute mark they radioed for McKnight's twelve-vehicle convoy to make its way to their location. Having departed from the Mogadishu Airport upon lift-off of the helicopters, the convoy had taken twelve minutes to negotiate the narrow streets of Mogadishu to arrive at its holding position behind the

Olympic Hotel, about 200 yards from the target. All had gone well so far, with the exception of the lead vehicle which, having made a wrong turn during the journey, ended up as the last vehicle.

By the time the ground convoy arrived at the target area, the intermittent Somali gunfire began to increase. At this point, other than the serious injuries sustained by Blackburn and an incident of near fratricide when a group of Rangers mistakenly fired on some Delta operatives on the target roof, who, in turn, were firing at a Somali sniper, the operation was progressing exactly as planned.

Overhead, though, the airborne commanders saw a different scene evolving as growing crowds of armed Somalis poured into the streets, erecting barricades and moving in the direction of the task force's objective clearly marked by the circling helicopters. The smoke from burning tires, rising black against the sky, signaled not only for the Somali gunmen to muster, but it also signaled the start to one of the most vicious firefights ever experienced by an American force since the days of Vietnam. Based on previous experience, the Somalis knew the routine, only now they were planning to change this story's ending.

Mogadishu was in chaos with masses of people everywhere, smoke rising, and the sound of heavy gunfire echoing off buildings. Its inhabitants united against a common enemy—probably for the first time in decades—swarmed through the streets in search of vengeance as militiamen with megaphones shouted, "Come out and defend your homes." Somalis knowledgeable about American military tactics knew that a column of vehicles would have to extract any force fast-roped into the Bakara Market area. Thus, the preparation of roadblocks and ambushes were the orders of the day.

The ground convoy had moved from its position to pick up the prisoners in front of the building. Three blocks away, Eversmann radioed his platoon leader, Lieutenant Perino, requesting the dispatch of a Humvee to pick up the injured Ranger—Blackburn—who needed proper medical attention immediately for his internal wounds. When informed by the lieutenant that no vehicle could be dispatched because of the heavy fires they were starting to receive, Eversmann had Blackburn placed on a compact litter—stretcher—and directed two of his sergeants, Casey Joyce and Jeff McLaughlin, to assist Good and a Delta medic, Sergeant First Class Bart Bullock, in transporting the young private to the convoy.

With heads low and the four men posted on the handles at the corners, the litter team began to run down the street, Bullock had an IV in hand feeding the precious liquid into Blackburn's arm. On taking fire, the team put the litter down and returned fire every few steps. Their movement was too slow and very dangerous.

Realizing the need to speed things up, Joyce volunteered to secure a Humvee and began running down Hawlwadig Road toward the convoy. The first vehicle he encountered was a cargo Humvee with a group of Delta soldiers and Navy SEALs. A quick explanation from Joyce had the vehicle's driver back in the Humvee and racing a block up the road to where the remainder of Joyce's group was waiting.

With Blackburn safely aboard, the Humvee returned to the main convoy while the litter team ran back down the road to its chalk. Understanding the gravity of the injured Ranger's condition, McKnight authorized the SEALs' vehicle to transport Blackburn to the American compound. Since the vehicle was unarmed and relatively unarmored, McKnight ordered Staff Sergeant Jeff Struecker and his two armored Humvees, one armed with a .50-caliber machine gun and the other with a Mark 19 40-mm automatic grenade launcher, to escort the SEAL vehicle.

Struecker's had originally been the lead vehicle that had taken a wrong turn. This time he knew where he was going and had mapped out a simple route. But the Somalis were one step ahead. Barricades had been erected and roadblocks emplaced. Armed street fighters roamed within crowds of civilians, masking their presence behind these supposed non-combatants until they opened fire. The intensity of the incoming fire increased as the little three-vehicle convoy pressed on, bypassing obstacles or ramming through them, shots zipped at them from every barricade, from every street and alley, and from windows and rooftops.

In the lead vehicle, Struecker directed Sergeant Dominick Pilla, his M-60 machine gunner, to fire upon targets to their right while the .50-cal concentrated to the left. Shortly thereafter, a simultaneous exchange between Pilla and a gunman resulted in a bullet to the sergeant's forehead, striking just below his Kevlar helmet, killing him instantly, splattering blood and brain matter about the vehicle.

Demands for a situation report (SITREP) on Pilla's condition finally forced Struecker to transmit over the radio the words he did not want to acknowledge: "He's dead!" With that declaration, all radio traffic ceased for thirty long seconds as the confirmation of the task force's first KIA began to sink in.

The three vehicles continued to move, albeit very slowly, for despite the intensity of the firing, there was a swarm of Somalis in the road. Relatively harmless flash-bang (concussion) grenades were tossed by the Rangers in front of the vehicles and the .50-cal fired overhead in only moderately successful attempts to disperse the crowds. When a slow-moving pickup truck loaded with people hanging all over it refused to move aside or speed up, Struecker had his driver push it off the road. Dazed, bloodied, and twenty minutes after their departure from the target area, the little convoy finally passed through the gates of the American compound.

Back at the target site, the situation was quickly worsening with all elements, particularly those on the perimeter, taking heavy fire. Incredibly, crowds of locals were still making their way through the streets, complicating the efforts of the Rangers and Delta teams to adhere to their restrictive rules of engagement (ROE) that prevented them from firing on seemingly unarmed and non-combatants. Such restraints were proving exceptionally difficult to follow as armed men jumped out from among the crowds, firing their weapons, and then jumping back into the security of the masses while others hid behind women, firing their weapons either from between a woman's legs or from under her arms. Other gunmen fired from the prone position, children in front and sitting on top of them, serving as human shields.

Minutes later, though, there were even more serious concerns as a call from Chief Warrant Officer 4 Clifton P. Wolcott, the pilot of the lead assault Black Hawk, designated Super 61, reverberated from the Task Force's speakers.

"Six-One's going down. Six-One's going down."

Within moments of this frantic call, the task force's nightmare began as rifles, hand grenades, and RPGs were fired or thrown from nearly every building window, doorway, rooftop, and corner in sight. During the course of Task Force Ranger's four-month campaign, Aidid's militia recognized that the U.S. task force's most significant asset was speed and mobility as represented by the frequent employment of the UH-60 Black Hawk helicopter. Eventually the militia developed a plan to negate this advantage.

The raid of 3 October, deep within their own territory, provided the Somali militia the opportunity to implement a tactic employing massed firepower in the form of rocket-propelled grenades, accurate at 300 yards against moving targets like low-flying helicopters. U.S. commanders had considered this tactic a possibility and a detailed, complex "downed aircraft" drill had been conducted just three days after the shoot down of a 10th Mountain Infantry Division Black Hawk. While a number of problems were worked out during drill, the task force clearly realized the need to rehearse and revamp CSAR . . . Combat Search and Rescue . . . actions in the future.

However, despite the success of an RPG against an airborne target, it was still believed to be a relatively innocuous anti-helicopter defense system. Since not a single TF 160 helicopter had been hit by the Somalis up to now, by either a bullet or an RPG, the belief in the Night Stalker's invincibility had only increased. In a morbid sense, it is ironic that Walcott's Super 61 Black Hawk had served as the "downed" aircraft during the CSAR drill.

Mobilizing quickly, SNA militia forces rapidly converged on the vicinity of the Olympic Hotel with a huge arsenal of RPGs they had been stockpiling for months—a significant failure of U.S. intelligence to detect. The militia quickly put their plan into effect. Around 4:10 PM, swarms of RPGs flew skyward. In less than ten minutes, one Black Hawk alone was the target of ten to fifteen RPGs.

Finally, at 4:20 PM, forty-eight minutes into the raid, the Somali tactic paid off. Having safely inserted his chalk of Rangers earlier, Black Hawk Super 61, piloted by Wolcott and co-piloted by Chief Warrant Officer 3 Donovan L. Briley, was hovering 75 feet above ground level (AGL) and searching for Somalis who could threaten the objective, when it was hit in the tail rotor by an RPG round fired by a gunner kneeling in an alley below.

Three blocks west of the crash site, the Rangers of Chalk Two in the northeast corner of the objective watched in stunned disbelief as a puff of smoke appeared near the Super 61's tail rotor and the aircraft began to shake violently. The blades began a difficult counter-rotation as Wolcott struggled to keep the seriously crippled eight-ton bird under control. Approximately 300 yards northeast of the task force's assault objective—about four city blocks—Super 61 crashed on the roof of a house located within a walled compound. The Black Hawk fell to earth on its left side, its top wedged against the remains of a wall in a narrow alley, its nose to the ground. Within, Wolcott and Briley lay dead, and the five others aboard—two crew chiefs, a Delta medic and two Delta snipers—lay injured.

Having previously rehearsed the possibility of an aircraft going down, the task force quickly implemented three contingency plans: provide cover with a nearby CSAR Black Hawk (Super 68), redirect the main body of Task Force Ranger from the objective to the crash site, and alert the Quick Reaction Force from the 10th Mountain Division to deploy from its location at the Somali National University to the Mogadishu Airport from where it could launch to support CSAR missions.

On the ground at the objective, Lieutenant DiTomasso, whose Chalk Two element was closest to the crash site, immediately radioed Captain Steele, requesting permission to depart the perimeter and move to the location of the downed Black Hawk. Steele, unable to break through the radio traffic to higher headquarters to make himself heard, ordered his platoon leader to proceed to the downed Black Hawk with half his force, leaving behind seven men, including an M-60 machine gun crew and an M-203 grenade launcher under the command of Sergeant Ed Yurek to continue perimeter security of the objective. DiTomasso had to run to catch up with his men, who had already begun to execute his orders by setting out at a trot without him.

Meanwhile at the Super 61 crash site, two dazed survivors quickly removed themselves from the wreckage. The second, Sergeant Jim Smith—one of the Delta snipers—emerged from the shattered aircraft armed with a rifle. Immediately moving about to defend the crash site, Smith quickly shot and killed one Somali gunman and set about engaging a group of others while taking a bullet in his left shoulder. Though wounded, the sniper was able to return to the downed Black Hawk to extract his injured partner, Staff Sergeant Daniel Busch. Smith then positioned Busch with his SAW and .45-caliber pistol against a wall only to see his friend suffer a serious wound to the abdomen, just below the plate of his bulletproof vest, as they continued to valiantly defend their position.

The Star 41 MH-6 Little Bird helicopter piloted by Chief Warrant Officer 3 Karl Maier courageously set down on the five to eight degree slope of a nearby alley called Freedom Road that was so narrow the rotor blades barely cleared the walls of two stone houses. Smith appeared next to the window of the Little Bird's copilot, Chief Warrant Officer 4 Keith Jones, mouthing the words, "I need help," for he was unable to be heard above the turbulence of the rotary blades overhead. While Maier held the controls in one hand and provided covering fire with his personal weapon in the other, Jones ran after Smith to the downed Super 61.

Within moments of Jones' departure, Lieutenant DiTomasso was nearly shot by a startled Maier when the Ranger turned the corner of a building to find the grounded Little Bird before him. With the lieutenant were his men, each having just completed the terrifying run of over three blocks, bullets boring down the alleys from every direction.

DiTomasso sent his small patrol to secure the downed helicopter's perimeter and to assist Smith and Jones, who were dragging and carrying the critically wounded Staff Sergeant Busch uphill to Star 41. Opening a small back door in the aircraft, Jones placed Busch inboard and then assisted Smith inside. With Jones once again seated, Maier applied power and climbed for a quick flight back to the American compound. For Busch, though, it would not be quick enough. His wound was fatal.

Eight minutes after Walcott's Super 61 going down, the CSAR Black Hawk, carrying fifteen members of a highly trained combat search-and-rescue unit composed of Delta—including a captain and sergeant's major—and Air Force Para Rescue, commonly referred to as PJs (Para Jumpers), was hovering over Freedom Road, approximately seventy yards around the corner from Wolcott's Black Hawk. By the book and with his arrival, the CSAR commander would be in charge at the site of any downed aircraft.

Fourteen of the men had fast-roped in before the fifteenth man, Air Force PJ Technical Sergeant Tim Wilkinson, noticed that the team had forgotten to take its two medical kits. Waiting until the last man had exited the ropes on his side thirty feet below, Wilkinson tossed the kits out the door and jumped for the rope.

Exposed too long as a result of the delay, the CSAR bird took an RPG round on the left side, behind one of its two engines. The whirling blades overhead whistled, damaged by shrapnel. The pilot, Chief

Warrant Officer Dan Jollotta, radioed the command bird that they'd been hit and had to withdrawal from the action.

However, yelling on the aircraft's intercom, one of Jollota's crew chiefs bellowed at the pilot that there were still men on the rope. Dangling below was not only Wilkinson, but below him another team member, Master Sergeant Scott Fales. Fighting his natural instinct to extract the aircraft as it began to lurch from side to side, Jollotta eased the aircraft back into a hover, providing the few seconds necessary for the two men to complete their descent. This action undoubtedly spared both men from serious injuries and quite possibly death.

Peering through the swirling brown dust cloud generated by the rotary downdraft, the crew chiefs reported the two men clear. Ordering the ropes to be dropped, Jollotta eased back on his stick, pulling up and away from the insertion point. Trailing a thin gray haze of smoke, the mortally wounded aircraft started to make its way back toward the airfield three miles away. Fighting his controls the final mile of his approach, Jollotta was able to execute a rolling crash landing, hitting hard on the right landing gear but keeping the Black Hawk upright and intact. Having suffered no casualties, the crew immediately switched to a backup aircraft and headed back to the fight.

Meanwhile, hesitating for a moment on the ground to orient himself, Wilkinson gazed around the area, unable to spot either the crash wreckage or his teammates. What he did spot were the two medical kits still lying in the middle of the street. Snatching the kits up, he ran around the corner to the left and stumbled on the downed Black Hawk. His teammates, who had arrived only moments before, had already set up a tight defensive perimeter around the crash site along with DiTomasso and his small team.

The bodies of several Somalis were scattered nearby with others visible in the distance. Two Somalis strayed into the area and were cut down from the swarm of bullets buzzing about like hornets. The Somalis were careful, for the most part, and stayed outside of hand grenade range. Master Sergeant Fales, stretching up to peer inside the right side of the aircraft, was hit in the left calf and moved to the rear of the bird. Up front and on top of the wreckage was one of the aircraft's survivors, Delta medic Sergeant James McMahon, pulling Briley's body out of the cockpit.

Further examination of the aircraft's interior found the whole front-end of the Black Hawk folded onto the remains of Wolcott, trapping him in his seat from the waist down. The real problem now became the extraction of the body, for there appeared no easy way to reclaim it. Abandoning their aviation comrade was not an option for the Rangers or Delta.

As Wilkinson checked the interior of the wreckage searching for remaining crewmen, classified material, weapons, and any other sensitive equipment, he discovered the left side gunner Staff Sergeant Ray Dowdy still in his seat and buried under a large pile of debris. A Delta medic, Sergeant First Class Bob Marby, made his way into the wreckage to assist, but not before Wilkinson had freed the trapped gunner.

A fusillade of bullets ripped through the skin of the aircraft with the three men inside. All three men were hit—Wilkinson in the arm and face, Dowdy the tips of two fingers shot off, and Marby in the hand. Pulling up the Kevlar floor panels, Marby propped them up where the bullets had entered. Then, to avoid any additional exposure, the two medics tunneled out from the left troop door, bringing Dowdy out with them.

A casualty collection point was established around the bent tail of the aircraft where Fales had moved. Repeat trips back into the bird resulted in the bulletproof mats being brought out and placed around the collection point. Hunkering down behind the mats, Fales traded rounds with gunmen in the alley while the two medics patched up the three survivors of the crash—including Dowdy's right side partner—who were all now consolidated at the tail.

From the northern end of the Mogadishu Airport in the American compound, the soldiers and airmen stood in awe observing in the distance the massive fireworks display of rockets, RPGs, and tracers. Streaking across the sky, these lethal rounds sought out the helicopters, seeking another victim.

It did not take long for the situation to worsen dramatically when an RPG claimed a second helicopter less than twenty minutes after the first. Overhead, Chief Warrant Officer 3 Michael Durant in Super 64 had been directed to take Super 61's orbital spot over the target area opposite Super 62, piloted by Chief Warrant Officer Mike Goffena. Durant was in the process of making his fourth or fifth circular racetrack at rooftop level when his Black Hawk shuddered from the detonation of a warhead against the rear rotor. From behind Super 64, Goffena observed large chunks of material rupture from Durant's gearbox as oil spewed from the exposed engine rotor in a fine mist. Goffena quickly radioed Durant, informing him that he'd taken an RPG in the tail.

A scan of the instrument panel by Durant indicated that all readings still fell within equipment operating parameters. Knowing that the Black Hawk had been designed to fly without oil for a limited time, he felt confident that he could bring the crippled aircraft back to the American compound, which he could see in the distance—only a four-minute flight to the southwest. Ordered by Lieutenant Colonel Matthews in the C2 bird circling above to withdraw from the fight, Durant pulled out of his orbit and headed southwest with Goffena trailing in Super 62 for the first mile of the journey. Suddenly, just as Goffena was preparing to return to the fight, the rotor and three feet of the tail assembly of Durant's aircraft disintegrated. Super 64 began to plummet like a rock.

With the center of gravity shifted, the aircraft began to spin rapidly to the right, nose lunging down. The severity of the rotations intensified as the Black Hawk corkscrewed toward the ground. Durant struggled for control as his copilot, Chief Warrant Officer 4 Raymond A. Frank, raised his arms overhead to throw the switches on the craft's two engines in an attempt to cut power. Seconds before impact, Durant radioed, "Going in hard! Going down!" as the nose of the bird rose just enough to level off the aircraft, impacting on top of a frail shack. As the dust from the crash settled, Goffena was able to report while circling above that the aircraft had remained upright,

having come to rest amid a cluster of makeshift tin huts, thus increasing the odds that there would be survivors aboard. A low pass by Super 62 confirmed the movement of at least three of the four downed
airmen.

The time was 4:40 PM and Super 64 had crashed in a neighbor-
hood called Wadigley, about 1,500 yards southwest of Super 61's location. The task force's ultimate nightmare had been realized, and
the American command and control system was now stretched to
the breaking point. One bird down was bad enough but Task Force
Rangers' contingency plans could cover such an event. A second
Black Hawk down, though, had never been seriously considered.
Given that this improbability was a reality, the only task force elements available were already committed to battle. For the moment,
at least, Durant and his crew were on their own as Goffena watched
an ever increasing mob of Somalis close in for the kill.

But Goffena in Super 62 was not about to abandon his mates on
the ground. Inboard were his two crew chiefs manning the aircraft's
two mini-guns. In addition, there were three Delta snipers, Master
Sergeant Gary Gordon, Sergeant First Class Randy Shughart, and
Staff Sergeant Brad Hallings. On one run after another, Goffena
brought the Black Hawk down to the deck close to ground level on a
low sweep, employing his blade wash and door gunners to dispel the
gathering mob while the Delta snipers searched and eliminated any
identified RPG gunners. Small arms rounds penetrated the thin skin
of the aircraft, increasing in intensity with each gun run.

Additional aircraft arrived to assist Goffena with the close air support and crowd suppression. Repeated calls by the pilots for ground
assistance were met by repeated assurances that a hastily assembled
ground convoy located only two miles away at the American compound would soon be on its way.

Having been directed to load his two dozen prisoners aboard
one of the five-ton trucks and consolidate his forces at the Super 61
crash site, McKnight's nine-vehicle convoy had not even begun to
move before Durant's Black Hawk was shot down a mile south of

the convoy's position. New orders from the JOC directed McKnight to detour over to Wolcott's aircraft to provide what assistance they could and then move on to Durant's.

Incoming fire grew in intensity as the convoy struggled to find its way through the narrow, smoke filled and bullet-riddled streets of the city. Overhead, the command and control bird tried to direct the movement of the wheeled vehicles marked with large fluorescent-orange panels on top to aid and assist identification.

Delays in communications had deadly consequences. Commands from above to turn left or right were not received by McKnight until the lead elements of his convoy had passed the intended corner. Consequently, they were either consistently bypassing the right roads or turning onto the wrong ones. Another significant disadvantage was directing the fight from above. The airborne commanders could not really see the pounding the convoy was taking. What at first seemed a simple task—moving to the crash sites—eventually would prove, in reality, impossible to do.

Contrary to the initial plan of returning to the American compound following the securing of the captives, McKnight was directed to move his convoy in support of those troops located at the first crash site. Unfortunately, all hell broke loose before the vehicles even had begun to move when the convoy lost its first vehicle, a five-ton, to multiple RPG hits in front of the target building. Fortuitously, the vehicle was empty at the time, its crew having exited to establish local security.

Casualties began mounting steadily as the column pulled out and attempted to make its way through the battle torn streets to the first downed Black Hawk. Wounded were gathered and placed in the eight remaining vehicles—six Humvees and two five-ton trucks—during intermittent halts. Injured gunners manning the M-60 and .50-caliber machine guns and Mark 19 grenade launchers traded places to keep the heavy weapons operating. The convoy popped smoke to mark its location for the command above.

The airborne command and control elements directed McKnight

to take another route, which brought him right back to a location just a block north of the target house—the very same location where they had loaded the prisoners. That relocation proved most fortunate for Staff Sergeant Eversmann and the eleven other remaining men of Chalk Four. Pinned down by enemy fire since the start of the mission, Eversmann apparently received and acknowledged a directive from a Ranger officer to move out though he presumably knew he wouldn't be able to with only four or five men still capable of fighting. Cut off, unable to maneuver, and left behind, these Rangers were most fortunate, indeed, and Eversmann most certainly knew it as he loaded his battered and bloody men aboard the vehicles that had made it back inadvertently to their rescue.

With the members of Chalk Four safely mounted, the convoy continued to move, only to find at each dirt crossroad a virtual shooting gallery, forcing it to cross intersections one vehicle at a time. Every member of the convoy who could fire a weapon, including the drivers, engaged gunmen in the alleyways, on rooftops, and in windows.

Somalis were everywhere. A fifteen-man group of gunmen ran along a parallel road, pursuing the convoy and placing heavy fire at each intersection the Americans passed through. For Task Force Ranger, the rules of engagement were changing. For over an hour, the hard-fought battle had been waged, the sites, sounds, and smells of the conflict permeating the city. Forewarned, there could be only two types of people caught in the area of operation: fools or those desiring to be part of the fight. Both types presented serious problems for the Americans as supposed noncombatants began to participate actively or were seemingly willing to be used as human shields. In either case, they became the de facto enemy and, under the circumstances, were treated as such.

Back at the target house following McKnight's departure, those left behind with orders to make their way to the crash site were also encountering equally stiff resistance. The remainder of DiTomasso's Chalk Two was ordered to move first, and to make its way to the lieutenant's location. That was easier said than done. With the sounds of a raging firefight rising in the distance, the seven men began to move east along a 30-foot-wide alley. Warned to stay away from the walls, which served as funnels for bullets, the patrol moved down the center of the dirt road. The patrol began to take heavy fire as gunmen sprayed their automatic weapons from doorways, windows, and around corners. The Rangers laid suppressive fires north and south as they leapfrogged through each intersection.

Turning a corner three blocks later, the patrol, without having taken a casualty, linked up with the remainder of its chalk, which had established a small perimeter. DiTomasso was hunkered down behind a Volkswagen "Beetle" with the newly arrived patrol leader, Yurek, when the car began to rock from the impact of heavy caliber machine gun rounds fired from a large tripod-mounted gun located in an alley. Taking a LAW, Yurek fired, but his round fell short of the mark. Switching to his M-203 grenade launcher, his aim with this weapon was devastatingly accurate as the heavy machine gun and its crew were blown away by the explosion of the 40-mm round.

The men of the CSAR team watched as the Rangers from Chalk Two moved into position around the downed Black Hawk, relieved to know their numbers were growing, for the action around them was steadily increasing as rounds ricocheted all about. Some of the shots were coming from the vicinity of a clump of trees about twenty yards away. As one Ranger hosed the area down with 5.56-mm fire from his Squad Automatic Weapon (SAW), a CSAR member and some Rangers tossed hand grenades, one of which was tossed back after the others had exploded—its pin had been pulled but a safety clip that rested on the handle had not been "flicked" off by the American who had tossed it. The Somali throwing it back did not make the same mistake. The grenade's detonation seriously wounded the CSAR man.

Captain Steele had intended to move in an orderly manner from his location to the crash site in conjunction with McKnight's vehicular movement from the target building. With Chalk Two either already at the site or on the move, the Ranger company commander had directed Chalk One under the leadership of Lieutenant Perino to lead the movement through the alleys, followed by Delta and trailed by Chalk Three. Chalk Four on the far corner of the perimeter—and a greater distance away than originally planned—was ordered to close on the group—an order that was acknowledged by the chalk leader but much tougher to carry out. Miller's Delta operatives fell in with the Ranger movement.

Unit integrity did not last long, though, and began to disintegrate within two blocks of movement as Miller and his men, feeling the need to advance at a more rapid pace against the withering small-arms and RPG fire, began to strike out on their own, believing that staying alive meant moving more quickly through the streets and alleys. Confused, some of the Rangers moved from their assigned chalks to join the faster paced Delta movement.

It would appear that command and control, integral to the efficiency and effectiveness of any combat unit, had broken down. Indeed, loss of control was complete and should have been expected. For those who experience intensive close-quarters combat, such moments are nothing less than sheer terror, as so movingly displayed on the screen during the opening minutes of Steven Spielberg's movie, *Saving Private Ryan*.

In real-world combat, commanders can reasonably expect to exercise command and control prior to enemy contact and again on consolidation. As for the moments in between: even if a commander were able to issue orders during such a movement, he most likely would not be heard or—if he is—acknowledged, depending on the intensity of combat and the effectiveness of the enemy. Tempo and cohesion of the unit during such times are left to the initiative of NCOs—who stepped up exceptionally well during this engagement—and the individual.

During the 10th Anniversary of Operation Just Cause, General (Ret) William F. Kernan—former commander of the 75th Ranger Regiment—stated "When the battle was joined, when the first shot was fired, the success or failure of the fight was in the hands of my team leaders." In the desperate alleys of Mogadishu that day in October 1993, success and failure were in the hands of these NCOs whose initiatives and action were a clear reflection of their intelligence and training.

The force on foot moved east until meeting up with Freedom Road. A Delta team with some Rangers in tow, led by Sergeant First Class Paul Howe, were the first to round the corner of the wide dirt road and head north on a down slope to the crash site only two blocks away. It was a very long two blocks.

As the team rounded the corner, the first RPG fired at the group impacted on a wall close by, knocking some of the men off their feet and wounding one with shrapnel on his left side. Seeking shelter, Sergeant First Class Howe kicked in the door of a one-room house. The team filed in to treat the wounded comrade, reload, and recover for a moment from the laborious run with heavy equipment that had left the soldiers breathing hard and sweating profusely. A quick examination of the wounded man found him still fit to fight.

Trailing in the group behind Howe was Lieutenant Perino. Sergeant Mike Goodale was hit in the upper right leg, necessitating the need to move him off the street and into a courtyard as the bullets and RPGs flew along the road and alleys. Some Delta men soon joined them.

Captain Steele led the last group to turn onto Freedom Road, and he felt time was quickly running out to link up with the convoy ground unit—not knowing that McKnight and his men were being hammered and not waiting at the crash site as originally planned. The convoy, though, was not the captain's primary concern at the moment. He and his sixty Rangers were for the most part without a clue regarding the location of the remainder of the task force.

Steele and his radio-telephone operator (RTO), Sergeant Chris Atwater, were face down in the dirt, watching the last of the Deltas from the first group advance through the intersection. Suddenly, one of the men, Sergeant First Class Earl Fillmore, went slack, his black hockey-style helmet recoiling up as a bullet exited his head. He was dead before he hit the ground. The Delta following Fillmore stopped to grab the KIA but he, too, was hit in the neck after a few short steps. A third Delta assisted the wounded man with the recovery. Around the intersection they knew now they were in great peril.

Casualties continued to mount in the vehicles as the lost convoy fought its way through the city, the unarmored five-ton trunks, in particular, serving as large, inviting, and extremely vulnerable targets for RPG rounds. Periodically, the slow-moving convoy would be seen to the west by those at the Super 61 location, passing within a block or two of the crash site. Some of the convoy's drivers were also aware of their close proximity to the downed aircraft but made no effort to relay this information to McKnight, who was in the lead vehicle, a Humvee. This lack of communications was the result of command failure at every level from section leader on up. In their haste to leave the target site, few if any of the other members of the convoy had been informed of the change in plan to close on the crash site. Those who saw the downed Black Hawk had never realized until much later that they were looking at their new mission objective.

Up front, McKnight, having sustained wounds to his arm and neck, was desperately trying to follow the directions of those circling overhead, who were now endeavoring to steer the convoy clear of potential problem sites. Periodically, the column of vehicles would make a sharp left or right but to no avail. They never found the crash site; they were always under attack. On one occasion they were forced to make a U-turn based on instructions from those overhead on a narrow street—the five-tons having to perform a three- or four-point backing maneuver to turn around—returning through a storm of heavy fire they had just successfully run.

The vehicles of the convoy were proving to be more of a liability than an asset. Lightly armored and constantly forced to stop and take on casualties, the vehicles made inviting targets. Worse yet, the Rangers were not trained for such a task. They were dismounted light infantry warriors and had little experience with convoy procedures. Their inexperience continually showed as they stopped at intersections or just beyond, forcing those behind to remain exposed. At the halt, those who were still mobile dismounted the vehicle to secure the column, resulting in more casualties that had to be placed on the vehicles.

In the end, the Rangers were damned no matter what they attempted. They could not abandon the vehicles for they were loaded with their dead and wounded. Yet, when forced to halt for whatever reason, the convoy was so long that some vehicles would always be exposed and become virtual bullet magnets, forcing the occupants to dismount and establish local security, which would result in more casualties. It was a vicious cycle of death and horrendous wounds.

Manning the .50-cal on the convoy's trail Humvee was Sergeant Lorenzo Ruiz. Like many of the other Rangers who were originally assigned to the convoy, he had removed the Kevlar plates from his flak vest for a more comfortable fit in the vehicle's seat. During the battle, a bullet penetrated the unarmored nylon of the vest, entered his lower right chest and exited out the back. The young sergeant died of his wound.

Soon after Ruiz was gunned down, the convoy made another halt. Dismounting to secure their vehicle, Sergeant Casey Joyce and Specialists Aaron Hand and Eric Spalding quickly found themselves in a vicious firefight at the end of an alley. Sprinting from the trail vehicle, SEAL Petty Officer (PO) John Gay placed some suppressive fires down the alley, allowing the seriously exposed Hand to evade the enemy fires safely and make his way back to his vehicle.

Joyce was across the alley, behind cover, placing efficient and effective fires against the enemy. A short automatic weapon's burst fired down on the kneeling Ranger from a window above caught the unsuspecting sergeant in the back. Despite the heavy fires and the risk involved, Specialist Jim Telscher was able to make his way to Joyce and drag him back by his shirt and vest to the vehicles. A quick examination showed that a round had passed through the top upper back of the vest, where the vest had no armored plating, and passed through his torso, exiting his abdomen. Later, Joyce died of his wound.

Racing to the front of the stalled convoy, Rierson, the senior Delta NCO present, spoke briefly with McKnight. It was only then that the sergeant realized that they were trying to make it to the Super 61 crash site. Moving back down the column, Rierson conveyed that information to each vehicle as he passed. Reasonably, there was deep frustration for the men as they realized they would have to pass back through sites they'd already paid for with blood.

The order to press on with the mission was passed on from the JOC to McKnight along with an altered set of directions. Shortly thereafter, McKnight rounded a corner and encountered a roadblock. The column had to stop, and the men deployed to provide local security as a tremendous volume of fire inundated them. Two more Rangers went down from wounds and were dragged back to their vehicles.

Remounting, the Americans once again began to move just as they were hit with a barrage of RPG rounds. The second vehicle in line, a Humvee, caught one of the rounds as the warhead penetrated the light armored skin just in front of the gas cap. Detonating inside the vehicle, the round blew through some sandbags within the interior, ejecting Delta Master Sergeant Tim Martin, Specialist Telscher, and Private Adalberto Rodriguez from the rear of the still moving vehicle.

Black smoke filled the Hummer from small fires erupting inside as the vehicle came to a halt. Rodriguez, tossed approximately ten yards from the vehicle, was struggling to his feet when he was accidentally run over by the five-ton that had been following them, its driver momentarily dazed and disoriented by another RPG detonation. Martin had been leaning against the sandbags inside the vehicle and the main force of the blast had passed through his lower body, practically tearing the master sergeant in half. Soldiers again scrambled from their vehicles to secure the three men on the ground and the other wounded and stunned men still left in the destroyed Humvee. Seriously wounded, Rodriguez survived his injuries; Martin, though still alive, did not.

Farther back in the convoy, Private Ed Kallman, driving a Hummer, was waiting for another vehicle preceding him to clear an intersection when he caught sight of a smoke trail heading right for him. The RPG impacted against the vehicle's door just next to the young driver, the resulting blast knocking Kallman out. Regaining consciousness just a few seconds later, he resumed his driver's position and pressed the accelerator to the floor. The vehicle responded racing through the intersection. Fate had intervened on behalf of Kallman and all those aboard his vehicle, for the driver's window had been rolled down. The combination of the lightly armored Kevlar door and the reinforced glass window rolled down inside the door had proven to be enough to defeat the penetrating capability of the RPG.

With the attempt to deploy a second ground element to Durant's crash site from the American compound at the airfield—Staff Sergeant Struecker's convoy—the situation became even more confusing for McKnight as the command and control elements attempted to direct two independent ground forces to two independent crash sites. Repeatedly, the battalion commander would receive instructions directing him to the second site when, in reality, he was still searching for the first. Back in the column, the Delta medic, Sergeant First Class Don Hutchingson, was providing medical attention to those wounded while Rierson maintained communications with the Little Birds, relaying updates and calling in close air support.

Finding itself in a relatively safe environment for the moment, McKnight halted the convoy and reviewed its situation with some of the column's senior NCOs. A quick SITREP to Romeo 64 brought an insistent response pressing McKnight to move to the first crash site. But McKnight knew that the situation for his men was worse at that moment than the situation they probably would find at the location of the downed bird. He already had more dead and wounded with him than the other two crash sites combined, and if he did not get his most serious casualties back, there would be more dead. What vehicles he had left were barely running, and he was low on ammunition.

In addition to all that, he was still lugging around the prisoners. Though one had been shot dead and a second one wounded, the Somalis obviously were being very careful and not intentionally firing on the captive's vehicle. The convoy had made a gallant, though

fruitless, effort to assist the downed comrades. Frustrated and enraged at his inability to find the crash site, McKnight realized it was
now time to return to base.

Seven seriously battered vehicles remained: five Humvees and
two five-tons. McKnight's Humvee still led, followed by a second
Humvee—a cargo vehicle lacking heavy gun capability, dragging
its rear axle and being pushed along by one of the five-tons. Gay's
Humvee was running on three flat tires, while a fourth, Kallman's,
had all four flat, in addition to a large hole in the driver's door from
the RPG.

Encountering ambush after ambush, the soldiers were fighting
for survival now. At the hands of Specialist James Cavaco, a Mark
19 automatic grenade launcher atop a Humvee pumped round after round of deadly, accurate fire into second-story windows. As he
turned and began to engage targets down an alley to his left, he took
a round to the back of the head, just under his Kevlar, killing him
instantly. His body was pulled down and Delta Sergeant Paul Leonard stepped up to take the vacated position. Minutes later, a bullet
passed through the vehicle, striking Leonard in the left leg and taking off much of it behind and below the knee. With a tourniquet in
place, the resolute sergeant remained in the turret and continued to
place exceptionally accurate and devastating fires on the enemy.

Another gunner, Private First Class Tory Carlson, manning a .50-
cal machine gun, was also hit in the leg as he put down withering fire.
Wounded in the right knee, he trembled with fear at first and found it
difficult to breathe. Then, realizing it was not he but his buddies who
mattered, the fear passed, and he began firing at the enemy again.

In the second of two five-tons, Private First Class Richard Kowalewski, teasingly referred to as "Alphabet," continued to drive despite a wound to the shoulder. Private First Class Clay Othic was
struggling within the confined space of the vehicle's cab to apply a
pressure dressing to Kowalewski's wound when an RPG round entered the vehicle from the left side, severing the driver's left arm and
embedding itself in his chest, killing the young Ranger instantly. Failing to detonate, the two-foot-long rocket with fins protruded from
under the stump of the missing appendage.

Continuing to move, Kowalewski's now driverless truck rear-
ended the vehicle in front, the remaining five-ton that was carrying
the precious cargo, causing that truck to veer into a wall. Again, the
convoy halted to assess the situation and to gather casualties. Incredibly, Othic and Hand, the other man in the cab, suffered only minor
injuries. Kowalewski's body was moved to the rear of the vehicle with
the RPG round still embedded in his body. Hand recovered the missing arm. Not sure what to do, he placed it inside the cargo pockets of
his pants, sensing it would be wrong to leave it behind. The convoy
started again with a new driver in Kowalewski's vehicle.

The group finally encountered something they recognized, a
four-lane road that led to the K-4 traffic circle and their home base.
Gay's Humvee was now the lead vehicle with one Ranger KIA and
eight wounded in back. A wounded Delta sergeant was in the front,
stretched out between the SEAL and his driver. Encountering the
last of what seemed to have been an infinite number of roadblocks,
this final one was constructed of fifty-five-gallon fuel drums that had
been placed across the road amid other debris and set ablaze.

Fearful that his vehicle would not be able to start again should
he stop—having been hit by dozens of rounds and running on its tire
rims . . . which the vehicle is designed to do but certainly not under
these most adverse of circumstances—Gay ordered his driver, fellow Navy SEAL PO Homer Nearpass, to ram the roadblock. Their
luck held as they breached the obstacle, going over and through the
flames. Behind, the remainder of the convoy followed as Gay led
them to the K-4 circle.

At the Mogadishu airfield, American soldiers, sailors, and airmen had gathered, listening to the firefight on the radio, aware of the
downing of both Wolcott's and Durant's Black Hawks. Some were
eager to move to the sound of the guns, while others felt a great deal
of trepidation and anguish, especially those who had already braved
the violent streets of Mogadishu once, and had lived to tell about it.
Such overwhelming fear was not at all unreasonable given the circumstances these men had faced. If courage is defined as action in

the face of that fear, then everyone in the fray was courageous that day.

Word soon spread that Staff Sergeant Struecker's two shot-up gun vehicles along with an additional two Humvees and three five-ton flatbed trucks were to make an attempt to reach Durant's crash site. As the rescue team of Rangers, Delta, cooks, clerks, and other volunteers began to assemble, a Delta sergeant took Struecker aside, offering some sage advice about washing down his blood-splattered vehicle. "If you don't, your guys are going to get more messed up. They're going to get sick."

One Ranger in particular, Specialist Dale Sizemore, watched Struecker and his men prepare to depart. Dressed in a T-shirt and athletic shorts, the young Ranger had injured his elbow a few days prior. With a cast on his arm, he was not only unable to deploy on the day's raid, he was manifested to depart on a flight for the United States later that evening. Informing Struecker that he was going to join the convoy, Sizemore ran back into the hangar, putting on his BDUs while searching for stray gear such as Kevlar and flak vest. Grabbing his SAW, the specialist returned to the vehicles, his pockets bulging with extra ammunition, his shirt unbuttoned and boots unlaced. In response to Sergeant Raleigh Cash's decision not to permit Sizemore to participate as long as he wore a cast, Sizemore cut it off.

When confirmation was received that McKnight and his men were returning to the compound, the hastily assembled convoy of men and vehicles began to deploy to Durant's location less than two miles away. Taking a left turn out of the compound's back gate, Staff Sergeant Struecker's lead vehicle had traveled no farther than eighty yards when all hell broke loose as the seven vehicle convoy—two Hummers in the front, two to the rear, and the three five-tons in the center—came under heavy fire. Leading the convoy, Struecker's Humvee was the primary target as an RPG round scraped across the top of his vehicle, continuing on to detonate against a concrete wall just a short distance away. The force of the explosion raised one side of the heavy vehicle clear off the ground.

Struecker very quickly realized that continued movement forward along this route would invite disaster, especially for those lying exposed on the flatbeds of the five-tons. A quick order to his driver had the vehicle backing into the second vehicle in column, which backed into the third. This sequence of events was repeated until all the vehicles had reversed back into the safety of the compound.

Seeking another route from the C2 helicopter circling above, Struecker led his small relief column back out the gate, trying a right rather than a left turn this time, and encountered a large roadblock composed of wire, dirt, and large chunks of concrete. The .50-cals in the convoy opened up, laying down suppressive fires. Beyond the roadblock, Struecker could see a concrete wall that surrounded the immense ghetto area in which Durant's Black Hawk had fallen.

A quick assessment of the situation virtually eliminated all his options at this location. While his Hummers could traverse the roadblock, the five-tons could not, and even if he were able to get his force on the other side of the obstacle, the wall itself served as an additional barrier with no points of passage in immediate view.

In response to Struecker's request for a third route, the helicopters overhead could find none readily apparent. Unwilling to concede defeat, the sergeant continued to press for an alternative when word came that the only option left available was for the convoy to make its way around the city and come in from the rear. Struecker seized the option.

Soldiers dismounted, providing total peripheral local security and engaging targets as the vehicles—the five-tons in particular—struggled to turnaround in the narrow street, ramming into walls as they moved back and forth. With the convoy finally headed in the new direction, the security elements remounted and the trucks began to make their way along a road that ran southwest through the city, with only a few bursts of fire directed at them. From one rise on the road, they could actually see Durant's crash site only a quarter of a mile away—not knowing that five of the valiant defenders, including two Delta snipers inserted earlier—were already dead and that the sixth, Durant, having suffered a serious beating at the hands of an angry mob, was now a captive.

They continued on their route, eventually arriving at the K-4 traffic circle. The scene they came upon at K-4 was shocking. Before them were the smoking and bleeding remnants of McKnight's convoy that had arrived at the circle just moments before, the broken-down cargo Hummer still being pushed by a five-ton. Struecker and his men were even more stunned to find the shot-up vehicles overloaded with dead and wounded Rangers.

For McKnight and his men, who had expected a serious fight at the circle, the sight of Struecker's relief column approaching them was a godsend. Struecker and his men quickly encircled the battered force, setting up security. The transfer of the dead and wounded from the most heavily damaged vehicles began as the cargo Humvee and Gay's vehicle were abandoned and destroyed.

In the midst of the transfer, there was a bright flash followed by an explosion as an RPG detonated under one of Struecker's Humvees. The force of the explosion blew the vehicle into the air but it landed back on all four wheels, undamaged and still fully functional as demonstrated by its .50-cal, which quickly eliminated a Somali sniper in a tree.

Though it had less than a mile to travel, the combined force, with Struecker's element pulling rear security, exchanged fire with everything it encountered along the way, pouring heavy volumes of fire in all directions and down every alleyway. Then, as though passing through a curtain, a stillness settled on the convoy just a few blocks from the compound. The Somali guns fell silent as large crowds gathered through the open-air markets.

The masses parted, allowing the embattled convoy slowly to make its way through the gate of the compound. In a surreal moment, as the column traversed the final one hundred yards, the Somalis all turned to face the Americans and proceeded to applaud; some applauding for the American presence, others applauding for their being bloodied.

The time was approximately 5:30 PM; McKnight and his men were finally secure—though, unfortunately, one of the convoy's survivors,

First Class Matt Ricrson, was killed a short time later by a mortar attack against the airfield that also wounded twelve others. Within the compound, personnel surrounded the vehicles to find wounded Rangers piled on top of dead ones, who were on top of the original twenty-four Somali prisoners still alive, who were on top of Somali dead.

Some of the dead Rangers' eyes were wide open, the fear still somehow captured in them. Others had a look of peace. Kowalewski still had the unexploded RPG round stuck in his chest. The recovery detail carefully removed his body and built a bunker around it as a precaution against the warhead detonating.

The convoy's casualties had been severe: four KIA, numerous wounded, and three dead prisoners. Yet the battle was still in full swing and many more casualties would result before it was all over. For the moment, at least, the ninety-nine men scattered in groups around the site of Super 61 were on their own.

Hours earlier, the air commander had denied requests from his four MH-6 Little Bird copilots that they be inserted on the ground to defend Durant and his three crewmen. The air commander also denied two additional requests from the Delta snipers, Gordon and Shughart, to be inserted on Goffena's Black Hawk; however, once he discovered that Struecker's relief convoy was turned back, he approved the third request by the two NCOs.

Upon hearing the news, Master Sergeant Gordon, the team chief, expressed his satisfaction with the decision by a smile and a thumbs up. Moving to the rear of the aircraft Gordon and Shugart set about making plans. Though the purpose of the insertion was to allow the two men—each armed with only his sniper rifle and a pistol—to provide first aid, establish a defensive perimeter, and secure the site until the arrival of a rescue force, all concerned knew that death was awaiting the two Delta NCOs below in the streets of Mogadishu. No rescue force would arrive in time before the growing number of enemy personnel observed closing in on the crash site overwhelmed them.

An eyewitness later stated, "Anyone in their right mind wouldn't have done what they did."

But Gordon and Shughart also knew that the four wounded men below would have no chance of survival without additional support. At a later ceremony, Master Sergeant Gordon's widow, Carmen, spoke of why she believed her husband acted thus. "Gary went back to save his fellow soldiers, not to die there. Gary was one hundred percent Ranger. He lived the Rangers' creed every day. He knew that he had a chance. He and Shughart wouldn't ever have gone out there trying to be heroes."

The first insertion failed as debris and small arms fire made a landing difficult. Finding a second site in a small clearing approximately one hundred yards from Super 64, Goffena employed the downward force of the blade wash—the wind-like effects of the large, whirling blades—of his Black Hawk as it hovered five feet above the ground to knock down a fence. Gordon tripped and fell as he ran for cover. Shughart, in his haste to disembark, forgot to disconnect his safety line and had to be cut free of the aircraft as it began to ascend.

The swirling debris, noise, and confusion of combat disoriented the two snipers. Crouched in the open field, Shughart motioned to Goffena expressing confusion as to which direction to move. The pilot brought his aircraft back down, leaned out the window, and pointed the way to Super 64 as one of his crew chiefs tossed a smoke grenade in the direction of Durant's bird. The last sight of the two intrepid soldiers as Super 62 lifted off to hover overhead with covering fires was of both men signaling a thumbs up as they began to fight their way under intense small arms fire through a dense maze of shacks to the downed Black Hawk.

In the wreckage of Super 64, all four crewmen had survived the crash. Durant, knocked out by the impact, regained consciousness to find the thigh bone of his right leg broken and a large sheet of tin punched through his shattered windshield and draped over him. Frank had his left lower leg broken. Both pilots had sustained back injuries. Unable to move, Durant secured his German MP-5K 9-mm rifle and prepared to defend himself from his seat as the copilot crawled from the wreckage out the opposite side.

Just as Frank moved out of his view, Durant was surprised and relieved by the arrival of Gordon and Shughart. Undoubtedly, a rescue team had arrived and their trial by fire would soon be over. Calmly reaching in, the two Delta men gently lifted and carried the injured pilot outside of the right side of the aircraft to a nearby tree. Behind him, the front end of his aircraft was wedged tightly against a tin wall, which the pilot covered with his weapon. Staff Sergeant William Cleveland, nearly comatose and covered in blood from the waist down was placed near Durant.

Gordon and Shughart moved to the left side of the chopper to extract the remaining crew chief, Staff Sergeant Thomas J. Field. Frank, having exited the cockpit from his left seat, joined the two Delta sergeants engaging the approaching militia and defending the exposed side of the downed helicopter.

Unknown to any of the six men at the site, Maier and Jones were once again on the ground with their Little Bird, personal weapons drawn, only about one hundred yards or so away. Having done what they could for Wolcott and his crew at the first crash site, they had landed to see what they could do for Durant and his men. Goffena circled above, observing Gordon and Shughart moving about the site and realized that the two Delta men would not be able to move the wounded men the distance necessary to link up with the Little Bird helicopter Star 41. Reluctantly, after a five-minute wait and a brief by Goffena of the crew's condition, Maier and Jones, fuel running precariously low, were forced to lift off to refuel.

Fate intervened again twenty minutes into Super 64's fight when Super 62 took an RPG round in the cockpit that knocked out the copilot and amputated the door gunner's leg. With the windshield knocked out, the right side of the aircraft blown apart, and the number-two engine destroyed, Goffena was still able to miraculously nurse Super 62 back in the direction of the airfield. Unable to make it to the flight line, he skillfully conducted a controlled crash landing in the dock areas, undoubtedly saving the lives of his crew.

With their air cover gone, Gordon and Shughart were on their own, facing an overwhelming number of Somali militia advancing on the wreckage of Super 64. Automatic weapons of the defending Americans covered all approaches to the downed aircraft and plenty of Somali bodies littered the area.

But time—and luck—were soon to run out on the gallant defenders. An exchange of gunfire brought a shout of anger and pain from Shughart on the far side of the wreck. Durant never heard from him again.

Moments later, Gordon moved to the right side of the aircraft, searching for ammunition and asking the dazed, confused, and painfully wounded pilot if there were any weapons inboard. Searching the interior, Gordon returned with the crew chiefs' M-16s in hand.

Reality—and probably a sense of hopelessness—finally struck the wounded warrant officer when the sergeant asked him what the support frequency was on the survival radio. With a sickening and nauseous feeling, Durant realized that such a question only meant one thing: the two Delta men had arrived at the site on their own, with no other support. There were no other rescue team members!

Following Durant's brief explanation of procedure, Gordon established radio contact, requesting immediate help. The reply, as it had been before his insertion, was that a reaction force was en route to their location. With that, the Delta sniper gathered his weapons and moved back around to the left side of the aircraft to engage the advancing militia.

Out of ammunition, Gordon returned once again to the wreckage, looking for anything to fight with, only to find very little. Gordon handed a loaded CAR-15 automatic rifle to Durant, whose own 9-mm weapon was either out of ammunition or jammed. Telling him "Good luck," Gordon made his way back to the far side of the aircraft armed only with a pistol.

At the nose end of the aircraft, Durant observed two Somalis trying to climb over. A short burst from his automatic rifle caused them to quickly disappear. Another Somali tried to crawl over the wall.

Durant shot him, as he did a second man trying to crawl around a corner. Off in the distance, less than a mile and a half to the south, Durant could hear the throaty roar of .50-caliber machine guns as Struecker's rescue convoy tried to deploy from the vicinity of the American compound.

Without warning, a hail of small arms fire on the left side of the Black Hawk, lasting for nearly two minutes, revealed a force of over a dozen concealed men focused their concentrated fires on the one remaining defender on the left side. Gordon's shout of pain was soon followed by silence.

The crowd surged across the clearing, pouncing on the four Americans who lay before them on the exposed side of the Black Hawk, one of whom was still alive, shouting and waving his arms as the mob grabbed his limbs, struggling to tear his and the other three bodies into pieces. Within a short time, the lifeless bodies of the Americans were being joyfully paraded and dragged naked through the streets.

On the far side of the aircraft, his weapon empty, and a loaded pistol strapped to his side but forgotten, Durant placed the rifle across his chest, folded his hands over it, and waited to die. The crowd rushed around the tail of the aircraft, assailing Durant and the body of his crew chief lying beside him.

High above, the cameras aboard the surveillance helicopters recorded images of "indigenous personnel moving all over the crash site." Nearly two hours after the aircraft had gone down, the fierce and deadly battle for Super 64 was over; all defenders and crew were dead with the exception of Chief Warrant Officer 3 Durant, whose life would be spared to serve as a hostage and whose eleven days of captivity, pain, and affliction were just commencing.

AUTHOR'S NOTE: Official records and their Medal of Honor citations reflect that Shughart was the first one killed in action with Gordon being killed just prior to Durant's capture. Subsequent interviews with Durant indicate that the official reports may have reversed the roles of the two Delta NCOs. Each Delta team member had a weapon designed and modified to his personal specifications—usually an M-14 or CAR-15. Based on the description of the weapon he was handed, it seems that Durant was provided Gordon's rifle as a final defense. It is doubtful the sniper would have handed over his own personal weapon and fought with another.

Durant had also indicated that he'd met Shughart before, during an air mission pre-brief, and he did not recognize the man who stood before him that day of the battle. Records, though, indicate it was Gordon who had attended the pre-brief in question, and thus it was Shughart he had not seen before. While the truth may never be known with certainty—though the audio tape of the final transmission from Super 64 would most probably indicate who died first—it is nearly irrelevant, for both of these courageous and professional soldiers fought and died as a team attempting to defend their fellow comrades in arms.

The casualties continued to mount in the vicinity of Super 61. Two Ranger gunners working an M-60 machine gun, Private First Class Peter Neathery and Vince Errico were each hit in the right arm. Lying on the ground behind a tin hut, Steele and another of his officers, Lieutenant James Lechner, were on the radio but did not notice Delta Sergeant Norm Hooten, who was trying to warn them from across the way when they began taking fire.

Steele was able to make it into the courtyard from which Hooten beckoned, and to which the previously wounded Goodale had been moved, but Lechner was not so lucky, suffering a bone-shattering wound to the leg as he tried to scramble from behind the shed. Delta medic Sergeant First Class Bart Bullock and a second man dashed out to bring the wounded man back to the courtyard for treatment.

On the radio again, Lieutenant Colonel Harrell in the command bird high above the fray—who was, by default, the ground commander as a result of McKnight's predicament—answered Steele's pleading demands for immediate support from McKnight's convoy and extraction.

"I understand you need to be extracted. I've done everything I can to get those vehicles to you, over," noted Harrell.

Steele's weary response was, "Roger, understand."

From the doorway of the courtyard, Steele motioned for more of his men to join him. Eventually, he ended up with a courtyard full of wounded men.

A block up the road, several small groups of Rangers and Delta had linked up with those defending the downed Black Hawk. They found themselves caught in a kill zone targeted by AK-47s and RPG rounds. Eventually an "L-shaped perimeter" was established that included the intersection of the crash at the angle of the "L." At the destroyed aircraft itself, the CSAR team continued to gather casualties and worked to recover the trapped body of the dead pilot. The Little Birds and Black Hawks provided air cover for the Americans, but the pilots' fires were restricted with the enemy mixed in among their fellow American comrades scattered about.

Sergeant First Class Howe was looking for a location to get his men off the street. With a second man's help, he broke through a security gate fronting a narrow courtyard between two houses in the vicinity of Lieutenant DiTomasso and his men. Inside one of the houses, they found a family cowering in the corner of a room. Following a quick search, their hands were tied, and they were moved to a side room.

Miller, who had finally caught up to the fast moving Howe, and the rest of the Delta team settled into the courtyard and houses. Miller, who had been monitoring the progress of McKnight's badly mauled column, knew it would be an extended wait until those at the crash site, now cut off and surrounded, were rescued.

Farther back along Freedom Road, the Rangers of Perino's Chalk One continued to move forward, creeping along the walls, searching for the position offering visibility without being observed, thus allowing them to eliminate targets with the minimum risk of exposure to return fire. Others, unable to find such positions, risked exchanging shots by exposing themselves. During one such exchange, Ranger Specialist Jamie Smith, standing behind a tin shed, took a round from a volley of gunfire that ripped through the shack. The bullet severed his femoral artery in the upper thigh. On the street, others lay wounded. Of Perino's original thirteen men, eight were casualties.

Lieutenant Perino and Sergeant First Class Kurt Schmid pulled the seriously wounded Smith into a compound off the street. An examination of the wound found a gaping hole in the Ranger's upper leg and blood everywhere. As Schmid began his treatment, the lieutenant radioed Steele to inform him that he could go no farther. "We have more wounded than I can carry," he informed his commander. Steele, though, wanted his unit concentrated in the vicinity of the wreckage and thus directed his subordinate to push on.

Perino was able to establish contact with DiTomasso, who was located across an alley from Super 61 in a stone house. Unable to confirm the position verbally, DiTomasso popped a red smoke grenade, revealing in the darkening sky that he was only fifty feet or so away from Perino.

Fifty feet or fifty miles. It was the same either way for Perino and his men. He did not have the resources to move the wounded through the crossfire to link up with DiTomasso. In response to Steele's insistence that he move forward to assist, Perino replied, "Look, sir, I've got three guys left, counting myself. How can I help him?" Faced with that circumstance, Steele relented, directing the lieutenant to occupy and defend the building at his current location.

It was nearly 5 PM and the fight had dragged on for an hour and fifteen minutes, during that time two Black Hawks had been shot down, two others had been seriously crippled, a lost and misdirected convoy was being severely mauled, and the dismounted rescue force was scattered about, itself now needing to be rescued. The situation only continued to grow progressively worse.

In Perino's courtyard, the medic had his hand in Smith's leg, trying to stem the flow of blood. Because the wound was so high on the leg, a tourniquet could not be applied. Clamps and hemostats could not be placed on the artery, either. The only option left was direct pressure with fingers placed within the leg, pressing directly against the ruptured artery.

Steele relayed Perino's request for a Medevac—a helicopter specifically designated to carry and treat wounded—to higher headquarters, though it was exceptionally difficult to hear anything in the cacophony of radio traffic. Headquarters' reply was of no help. The situation precluded any relief effort. Placing another helicopter in the caldron of the Super 61 area of operation was not an option, nor would it be at any time in the foreseeable future.

At the Super 61 crash site, the beat up, exhausted soldiers grew concerned as the sun set. Exceptionally well versed and trained as night fighters, the force found itself without its technological edge, having left night observation devices (NODs)—both for personnel and weapons—behind. Without that advantage, they all knew that whatever the night would bring, it could not be dealt with as effectively lacking the NODs.

As darkness fell, Wilkinson, the Air Force PJ Medic, received a radio transmission requesting his assistance with Private First Class Carlos Rodriguez, who had a round strike him in the buttock, pass through his pelvis, and exit from his upper thigh. Picking up his medical kit, he braved the fires in the open street, rushing from his place of relative security inside the downed Black Hawk to the courtyard across the street where Miller had set up his command post. With the use of wads of gauze and pressure applying pneumatic pants, he was able to stop the bleeding.

Wilkinson's only problem now was that the intravenous (IV) drip he'd just hooked up was his last. There were more, but they were back at the downed aircraft with his wounded partner, Fales. Back

out into the open he dashed to the Black Hawk. Gathering up a load
of the fluid, he once again defied the fires, making it through the
lead hail without being hit. A request by Wilkinson to have the seriously wounded Ranger extracted was met with a knowing look from
Miller. There was nothing more to be done. Maybe later.

After nightfall, the pace of the Somali attacks slackened notice-
ably until all grew quiet. Lights flickered in the city, indicating hum-
drum normalcy existed beyond the limits of the task force's defensive
perimeter. Fires were burning back in the vicinity of the target house.
The helicopter gunships were still in the air, as they would be all
night, periodically making runs when targets exposed themselves.
Later, unconfirmed reports would indicate that pilots and crewmen—under the supervision of the regiment's flight surgeons—were
given the option of taking amphetamines to keep them alert. Meanwhile, on the ground, ammunition was running low and many of
the men were parched, having left their canteens behind, believing
they'd be in and out, mission complete, within an hour.

Those across the street from Miller's compound were ordered
from the Command and Control (C2) helicopter overhead to fall back
through the intersection into the courtyard. During the withdrawal, Specialist John Stebbins was badly wounded by the explosion of
an RPG round that nearly took his head off. Later, two RPG rounds
struck the building occupied by Miller and his men, starting several
front-room fires. The smoke and fumes filling the house forced all
but the wounded and medics attending them outside for a time, before they reoccupied the structure.

In early evening, a Black Hawk made a daring run to deliver ammunition, water, and medical supplies into an area that had already
seen two helicopters shot down and two others hit and heavily damaged. Hovering over the intersection within the defenders' perimeter, the resupply helicopter took numerous small arms hits and deftly
avoided RPG fire as all supplies were pushed out. A Delta volunteer

crewman was added to the growing list of casualties with shrapnel wounds to the face and shoulders. The aircraft just barely made it out and back to the airfield, the fifth Black Hawk to be damaged or destroyed that day.

Within the area were ninety-nine men—nearly evenly split between Rangers and Delta—who had moved off the streets, hunkered down in four houses, the Black Hawk's wreckage, and associated courtyards. Farthest away was Steele with nearly thirty Rangers, some Delta, and nearly every automatic weapon that had moved to the downed Super 61 location.

At the crash site, Miller repeatedly requested Steele's assistance in consolidating their forces. Even though the two company commanders shared the same frequency modulation (FM) communications net, Steele never once spoke directly with Miller. When informed that Miller was unable to make contact with Steele by FM, one of the Delta team leaders at the Ranger's location offered his MX communications set to Steele to speak with Miller, only to have Steele refuse, obviously not wishing to explain to Miller why he felt he could not move.

Overhead, having heard Miller's requests, Harrell agreed that the force should consolidate on the crash site, thus providing greater defensive fires and making it easier for the gunships to do their job. Steele indicated he would make the attempt. However, upon hearing the news, Sergeant First Class Sean Watson, the leader of Chalk Three, told his commander he needed to reconsider. Five wounded men would have to be moved. Two were serious, each requiring a four-man litter team for transportation. The body of Fillmore had to be carried. Then there was the issue of just getting out of the door. On the other hand, the courtyard and building were easily defensible with the force they had. Upon reflection, Steele agreed with his senior NCO, informing Harrell that he would not be able to move with the casualties he had. Harrell left it to the two commanders on the ground; he was unwilling to make a decision either way.

After seeking communications and coordination with Steele for over two hours, Miller finally gave up. Still needing men to reinforce his perimeter, Miller ordered the Delta men with Steele to close on his position. Though aggravated, Steele did not attempt to stop the men. The Delta men lined up, and the first four rushed out the door into the darkness. The area lit up as weapons opened up from everywhere. Seconds later, the four men soared back through the door, landing in a tangled heap. Later, during early morning darkness while the un-disciplined Somalis were obviously inattentive, the Delta team was able to slip out and safely make its way to Miller's location.

At the Black Hawk, a hole was blown through the wall against which the wreckage rested to gain access to the stone house occupied by Miller on the far side, allowing all the wounded and dead—with the exception of Wolcott, whose body was still trapped in the heli-copter—to be transferred into the adjacent space.

Less than fifty yards away in another courtyard, Wilkinson, Perino, and others were trying to save Smith's life, each taking turns applying pressure to keep the punctured femoral artery closed. He needed immediate medical attention, and he was not going to get it lying in the dirt of the courtyard. Perino transmitted the request to Steele, who radioed Harrell. The Lieutenant Colonel's response was not encouraging, noting the earlier mauling of the resupply helicop-ter in less time than it would take for a grounded Medevac to load casualties.

Harrell pressed the issue with headquarters, though, inform-ing it that there would be more KIAs from WIAs if the two men in question, Smith and Rodriguez, were not extracted soon. Another helicopter would not be risked, command replied. Steele's two seri-ously wounded men would just have to wait until the QRF made it to their location, which they estimated would be within the next hour. Shortly thereafter, the news was broadcast over the command net that James Smith, the son of a retired Vietnam veteran Ranger Lieu-tenant Colonel, had bled to death from his wounds.

Close air support was stellar. Those on the ground knew full well they owed their lives to their aviation comrades; no one doubted they'd have been overwhelmed without their gun runs. Helicopter crews flew nearly nonstop for almost eighteen hours as Matthews rotated the AH-6Js and his remaining Black Hawks over the primary objective, breaking them off only to rearm and to refuel.

A five-word phrase was indelibly etched in the minds of all who listened on the radios: "Where do you want it?" To that question, the answer from Miller or the CSAR captain was usually to place it "Danger close"—as close as twenty feet from a friendly position.

Even when equipment failed, the crews quickly reacted. During one approach, a lead helicopter detected the launch of an RPG from a position 700 yards away. The pilot "sparkled"—illuminated the gunner with his laser and then passed off the target to another helicopter following behind. Accepting the target, the trail aircraft attempted to engage with its minigun, which malfunctioned. Quickly switching to rockets, the trail pilot killed the gunner. Overall time of engagement: three to five seconds.

During the course of the battle, the aircraft cycled through the Forward Arming and Refuel Point (FARP) up to a dozen times each. Armed during that time with eighty-seven rockets and 64,000 rounds of 7.62-mm minigun ammunition, the gunships would fire sixty-three rockets and 48,000 rounds into the engagement area. Over 12,500 gallons of fuel were pumped.

As the long dark night progressed, Miller and the CSAR captain continued to maintain their perimeter around the downed aircraft. When they were not working CAS suppressive fires, they were cross leveling—ensuring each man had his share of ammunition available—and running supplies between their positions. Working together, they had no issue as to who was in "command."

Support on the ground also continued to push in an effort to get support to the embattled Task Force Ranger. The QRF was C Company, 2nd Battalion, 14th Infantry Regiment (2/14 Infantry)—the "Golden Dragons." Having heard of the first Black Hawk being shot down, the ready company of the QRF received word at 4:29 PM to deploy to the airfield to reinforce Task Force Ranger. Needing to skirt around Mogadishu's SNA neighborhoods, C Company deployed in sixteen vehicles to the airport from their university location—which was two miles west and in the southern part of the city—by a long, secure route that skirted to the south, reaching the Ranger compound at 5:24 PM.

Again, another flaw in the Task Force Ranger plan became glaringly obvious at this point: Their planning was not integrated with the 2/14 Infantry mission. Garrison reported through the Commander-in-Chief Central Command to the National Command Authority (NCA). U.S. Major General Thomas M. Montgomery, who controlled the QRF, reported along a U.N. chain of command. Thus, there was no one involved with the QRF who was knowledgeable about the Task Force Ranger missions. Such a lack of integration had not been an issue during the first six missions, for the QRF was not needed. Now that it was, 2/14 Infantry needed to first deploy to the Rangers' compound at the airport to be briefed on its mission prior to heading out into the streets of Mogadishu. This was time that could ill afford to be lost.

Lieutenant Colonel Bill David, battalion commander of 2/14 Infantry, 10th Mountain Division, and the ground commander of the QRF—which also had elements of the 1/87 Infantry assigned to it—arrived with C Company and met with Major General Garrison. Attached to Task Force Ranger, David was directed by Garrison to rescue Durant's downed aircrew. David, given the enemy and the city he was to face, knew the effort required a force greater than what he had on hand. David's request to bring up the remainder of his battalion was not met with a great deal of enthusiasm. Time was of the essence and Garrison was hoping speed would substitute for David's lack of firepower. Little Bird and QRF Cobra gunship support were promised.

As David developed and issued his plan, a patchwork force of Task Force Ranger headquarters soldiers—cooks, clerks, staff—scrambled aboard six Humvees. Not quite sure where they were going, only knowing their comrades needed help, they preceded

David's convoy by a few minutes. David departed at 5:47 PM with his company in column with seven Kevlar-armored Hummers and nine sandbag-lined five-ton cargo trucks.

The six Ranger vehicles in front and sixteen QRF vehicles to the rear moved north at a good clip of thirty miles per hour despite the fire they had started to take as soon as they pulled out of the gate. Just north of the K-4 circle, at 5:54 PM, all that changed. Direct fires increased dramatically and a Ranger Humvee veered off the road and started to burn. A second Ranger vehicle was hit. The column slowed and began to close up on the lead vehicles.

Tracers were flying everywhere, as were RPGs, with estimates of up to 200 or more being fired during this segment of the engagement. The 150 men in the convoy dismounted. From the airfield, observers saw black smoke rising from the vicinity of K-4. The radio net grew eerily quiet. Though badly outnumbered, the Americans were very fortunate for the Somalis proved to be rather lousy marksmen, wounding only two of David's men.

The aircrews of Task Force 160 found it difficult to provide accurate fires for the force. Not accustomed to working with their type of aircraft and support, the QRF had to have it explained during the course of engagement how to mark and adjust for fires. Communication was reestablished at 6:15 PM when David, in frustration, reported that they could move no farther.

The QRF unit continued to fight until 6:21 PM, at which time they were ordered to withdraw. Even with the support of A Company 2/14 Infantry and the QRF Cobras, it took David nearly an hour to break contact after an expenditure of over 40,000 rounds of ammunition and hundreds of grenades. Clearly, a larger, stronger, and armored force would be required to break through to the trapped Rangers.

The QRF's second effort involved the use of a heavy armor force composed of a platoon of four Pakistani American-manufactured M-48 tanks and two companies of Malaysian mechanized infantry consisting of twenty-eight Russian-built BRDM wheeled armored personnel carriers (APC)—all painted the U.N. color of white. David augmented his original company with another 150 men from his A Company. An additional composite unit of approximately fifty Rangers in four Humvees under McKnight—ambulatory survivors from the lost convoy—Delta, and SEALs, joined the rescue force, which now totaled well over 425 men.

Humvees and five-tons were added to the convoy until it totaled seventy vehicles in all. David charitably called the entire procedure a "mess." Others called it a "three-ring circus." No matter what it was called, it was, in reality, the only rescue force in town.

The effort took over four hours to put together. In that blackout drive—special subdued lighting—and not normal white lights was to be used for the nighttime movement, operators had to be found who knew how to drive with Night Vision Goggles (NVGs). Moreover, while the U.N allied officers spoke English, their men did not, which resulted in delay as instructions had to be translated.

Then there was the issue of David having to explain to the Malaysians that it was their APCs he needed, not their soldiers. They would be replaced, with the exception of the driver, the track commander, an assistant, and a machine gunner, by his Alpha Company and whoever else could squeeze inside the rear of the vehicles—approximately nine Americans per carrier.

David's plan—developed on the hood of a Humvee, flashlight in hand—envisioned leading with the four Pakistani tanks, fighting their way to a midway point between the two crash sites, and then splitting his column in two directions. There was only one problem with the plan: The Pakistani tank commander had been directed by his headquarters not to lead the attack, just to support it.

The QRF deployed at 11:24 PM from New Port, a facility a few miles up the coast from Task Force Ranger's base; the site that had been used to affect the linkup between the multinational forces. As to who would lead, a compromise had been reached with the Pakistanis regarding the deployment of their tanks. Initially, they would lead along what became known as the "Mogadishu Mile" to the K-4 circle and a bit beyond, punching through any roadblocks or ambushes en-

countered along the way. Shortly thereafter, the APCs mounted with
the 10th Mountain's Alpha Company would pass through and lead
the remainder of the way.

The convoy was hit within five minutes of its departure by intermittent fires. Lieutenant Ben Matthews of 2/14 Infantry was riding in
the lead Pakistani tank. The ferocity of the ambushes increased dramatically as the two-mile long convoy neared the Black Sea neigh-
borhood, compelling Matthews to cajole the lead tank to continue on.
Behind the tanks, inside the unfamiliar APC vehicles, the American
light infantry could hear the rising storm of gunfire and explosions
outside, small arms fire pinging off the armored surfaces and RPG
rounds, hitting at the wrong angle to penetrate, deflecting off the
sloped armor. The convoy pressed on, firing thousands of rounds of
ammunition as it crashed through barriers and ambushes. Finally,
though, the lead tank would advance no more.

The two lead APCs moved forward and pressed the attack without any command and control element overhead to direct them. Disoriented and unsure of the route, the lead vehicle made a wrong turn
followed by the second APC. The lead APC was quickly hit in the
front by an RPG, bringing it to a burning halt. Inside the disabled vehicle, half a squad of infantrymen began to bail out as the second APC
came to a halt. Inside that vehicle, the dim, red lights went out, leaving the enclosed American infantry in the dark. The platoon leader,
Lieutenant Hollis, was aboard and ordered his men to dismount.

To make matters worse, the squad was stranded in a communication dead zone, unable to radio the main convoy that was now
passing them by to continue the rescue. The men blew a hole through
the wall of a nearby building and began to move in, but not before
Specialist Cornell Houston took a round in the leg.

The squad was trapped inside, caught in the midst of a serious
firefight with encircling gunmen. Houston was hit again, this time in
the chest by a sniper. This wound would prove to be fatal, causing
the specialist to die a few days later in an American medical facility
in Germany.

Outside, in the still smoking APC, a wounded Malaysian had
been left behind. Opting to take the mission on himself rather than

sending one of his other men, Corporal Richard Parent successfully made the run to the vehicle, dragging the wounded man back with him.

Somewhere in the city, a chained prisoner in a darkened room, Durant, could hear the progress of the QRF as the convoy battled its way forward—machine gun fire, Mark 19 grenade launchers, and rockets. At one point of the engagement, several APCs rolled by outside of his building—perhaps the two lost APCs? For a moment, he thought it might be a rescue attempt as an intense gunfight exploded outside. But those sounds faded quickly as the battle moved on.

With the two APCs heading in the wrong direction, the Pakistani tanks found themselves in the lead once again. Shrugging off RPGs and small-caliber rounds, the tankers were coerced to press on slowly, halting at times to fire. Approaching another flimsy barricade, the lead Pakistani tank driver refused to drive over it, thinking it was salted with antitank mines. Threats from the American lieutenant failed to budge him.

The convoy halted, and the American soldiers dismounted. The 10th Mountain deployed from the vehicles, establishing local security while others under heavy fire began to disassemble the obstacle with their bare hands. The convoy remained at a stop for over half an hour taking fire.

Incredibly, mobs of people were still making their way about, pressing to move closer. When warning shots overhead failed to disperse them, a few shots into the crowd would convince them of the necessity to do so.

Along a wall, Specialist Phil Lepre was preparing to move a few yards to get a better aim at a building from which shots were being fired. Ordering Private James Martin to move forward and place supporting fires from the position he was leaving, Lepre had only moved forward a few steps before Martin was hit and sent sprawling on his back. He had taken a bullet in the forehead. Martin would prove to be the final American soldier killed or mortally wounded in combat during that fateful battle.

At the release point following the obstacle breach, the convoy split into the two main 10th Mountain companies. A Company, commanded by Captain Drew Meyerowich, continued to move north until it was five hundred yards short of the American perimeter around Wolcott's crash site. C Company, under the command of Captain Michael Whetstone, moved west to the site of Durant's Black Hawk, but stopped just short of the location, unable to go any farther by vehicle because of the barricaded roads.

Delta Sergeant First Class John Macejunas gathered a small group of men and continued to press on by foot. A sweep of the wreckage showed there was nothing other than blood to be found of the men who had fought so valiantly there. Durant had been spirited away, and the remainder, though dead, were nowhere to be found. The group remained in the area long enough to ensure the destruction of the downed aircraft by well-placed thermite grenades. They returned with the remainder of its company, per David's directive, to secure Hollis and his men who were still trapped in the vicinity of their two destroyed APCs.

At the first crash site, Miller, Steele and their men could hear the heavy gunfire in the distance, the brilliant glow of the guns and explosions illuminating the black sky of the city to the south. The area around Super 61 had been relatively quiet for some time now. There were a few shots here and there and, periodically, a Somali would wander into the crash site only to be shot. Just to keep any thoughts of massing and rushing the site out of the enemy's head, Little Birds would swoop down from time to time, unleashing rockets and a streak of tracers from their miniguns.

The distant roar and thunder of the column advancing north intensified, its men firing in every direction, at everything, as they moved. The word soon came from command that the 10th Mountain Division would soon be there. The Rangers and Delta at the site pulled back from the windows, not wanting to become friendly-fire—fratri-

cide—casualties as the QRF stormed in. To minimize the possibility
of an accidental clash, flashing strobe lights were emplaced to identify their locations.

Steele's location on Freedom Road was the farthest south and
thus the closest to the advancing rescue force. Around 1:55 AM, Steele
heard the vehicles in the vanguard of the half mile-long column turn
onto Freedom Road. The images of dismounted soldiers soon appeared closing in on their building.

Steele's men called out, "Ranger! Ranger!"

"Tenth Mountain Division," came the reply in return. Linkup
had been accomplished.

The combined group still had four hours of work ahead of them as
they struggled to extract Wolcott's body from the crumbled remains
of his aircraft. A rescue saw that had been brought along specifically for the task was ineffective, its blade dulling on the Kevlar-lined
cockpit. Finally, the realization set in that they would literally have
to tear the helicopter apart. Chains were attached to the front of the
Black Hawk and a Humvee began to pull and separate the nose of
the aircraft.

As this four-hour contest of man against machine continued,
the wounded were assisted to the APCs parked in the center of the
road—an inviting target to any roving RPG gunner. Fortunately, the
enemy never seized the opportunity. Unable to see out of the vehicles
and unaware of the efforts taking place to recover the pilot's body,
some of the men began to grow restless as the wait intensified.

Finally, the recovery team's persistent efforts paid off, and the
warrant officer's body was freed. Gathering the body, the crew's personal weapons, one minigun, and classified documents, the force
torched the wreckage with some thermite grenades.

But the rescue column now had another problem: there were
too many people to ride in the limited number of vehicles available.
Space within each vehicle had been fully utilized by the force on the
way in. Now, on the way out, there were an additional ninety-nine
men to accommodate, some dead, many wounded.

Though exhausted, those who were still ambulatory in Task Force Ranger found that they had to remain dismounted, once again braving the fires as they made their way down the dawning streets of "Mog." The men departed the area the same way they had arrived—under intense fire. Running, leapfrogging through intersections, placing heavy fires down alleyways, the force only suffered one additional casualty, Sergeant Randy Ramagliea, as the main body fought its way toward their destination, the Pakistani compound that had been set up as a field hospital. A smaller group that had separated ended up in New Port.

The true horror of combat was readily apparent and on display at the Pakistani compound, a converted soccer stadium. Doctors, medics, and nurses performed triage on the mangled, brutalized, and bloody bodies that lay about covered in remnants of military uniforms. Men cried and hugged each other, glad to see a friend alive. Others cried at the sight or memory of a comrade killed in action. Some just sat, completely exhausted, numb, looking off with that "thousand-yard stare" many veterans talk of, but few had experienced until this day. Steele would later learn that a third of his company had been killed or wounded.

Though the unit had been seriously mauled, the fight was not yet out of Task Force Ranger. Many of the Delta operatives had departed the compound as soon as possible to make their way back to their hangar, where they were already preparing for any follow-on mission—cleaning weapons, restocking ammunition, and replacing lost and damaged equipment.

Even those Rangers who were badly wounded still had fight left in them. Specialist Rob Phipps lay on a cot in the field hospital badly battered and bruised, his back and legs heavily bandaged. Steele stood over the wounded man, placing his hand on the Ranger, telling him he would be all right. Recognizing his company commander through blood red eyes, Phipps reached up, grabbing the captain's arm.

"Sir, I'll be OK in a couple of days. Don't go back out without me."

Officially, the raid of 3 October—which came to be known by many of the participants as "Black" or "Bloody Sunday" and the "Battle of the Black Sea," and by those in Somalia as "Ma-alinti Rangers," "The Day of the Rangers"—came to a close around 7 AM on 4 October with the return of the final Task Force Ranger and QRF elements.

Given the savagery of the battle, "Task Force David" had suffered "moderate" American casualties: two KIA and twenty-two WIA. Malaysian casualties were one KIA and seven WIA while Pakistani losses were two WIA. Casualty figures were much more severe for the 120-man Task Force Ranger who suffered seventeen KIA—six of which were from B Company—fifty-seven WIA, and one MIA—Durant—during the course of the fifteen-hour firefight. All told, nineteen brave Americans were killed in action that day or would eventually succumb to their mortal wounds:

Chief Warrant Officer 4 Raymond Frank
> 160th Special Operations Aviation Regiment (SOAR)

Chief Warrant Officer 4 Clifton Wolcott
> 160th Special Operations Aviation Regiment (SOAR)

Chief Warrant Officer 3 Donovan Briley
> 160th Special Operations Aviation Regiment (SOAR)

Master Sergeant Gary Gordon
> 1st Special Forces Operational Detachment - Delta

Master Sergeant Timothy Martin
> 1st Special Forces Operational Detachment - Delta

Sergeant First Class Earl Fillmore
> 1st Special Forces Operational Detachment - Delta

*Sergeant First Class Matthew Rierson
> 1st Special Forces Operational Detachment - Delta

Sergeant First Class Randy Shughart
   1st Special Forces Operational Detachment - Delta
Staff Sergeant Daniel Busch
   1st Special Forces Operational Detachment - Delta
Staff Sergeant William Cleveland
   160th Special Operations Aviation Regiment (SOAR)
Staff Sergeant Thomas Field
   160th Special Operations Aviation Regiment (SOAR)
Sergeant Cornell Houston
   2/14th Infantry Regiment, 10th Mountain Division
Sergeant James Joyce
   3rd Ranger Battalion, 75th Ranger Regiment
Sergeant Lorenzo Ruiz
   3rd Ranger Battalion, 75th Ranger Regiment
Corporal James Cavaco
   3rd Ranger Battalion, 75th Ranger Regiment
Corporal James Smith
   3rd Ranger Battalion, 75th Ranger Regiment
Specialist Dominick Pilla
   3rd Ranger Battalion, 75th Ranger Regiment
Private First Class Richard Kowalewski
   3rd Ranger Battalion, 75th Ranger Regiment
Private First Class James Martin
   2/14th Infantry Regiment, 10th Mountain Division

*NOTE: Generally, losses for Task Force Ranger have been placed at eighteen KIA but do not include Sergeant First Class Matthew Rierson killed shortly after 3 October that led some to state a loss of nineteen KIA.

In comparison, Somali militia losses were staggering with United States and Somali sources placing them at nearly 500 killed—though a special U.S. envoy a few days after the battle believed the actual number to be closer to 1,000. The number of wounded was over 1,000. Certainly, though, all the casualties were not attributable to

American weapons, considering the number of RPGs—claimed to be more than 1,000—fired. How many Somali fatalities and casualties were the result of RPGs detonating outside of the engagement areas and of militia small arms fratricide? No one will ever know.

In 1997, former Chairman of the Joint Chiefs of Staff Colin L. Powell—an advisor who had sat in on all of the President's policy meetings on Somalia—stated in an interview, "Bad things happen in war. Nobody did anything wrong militarily in Mogadishu. They had a bad afternoon. No one expected a large number of soldiers to get killed. Is eighteen a large number? People didn't start noticing in Vietnam until it was 500 a week."

The world awoke on the morning of the 4th of October 1993 to a most barbaric sight: bodies of American soldiers being defiled. Broadcast by CNN, film footage of exuberant Somalis celebrating around and on the burnt hulk of Durant's Black Hawk gave way to a shot of a body, the remains of what was believed to be Durant's crew chief, Staff Sergeant Bill Cleveland, being dragged through the street at the end of a rope, the corpse kicked, poked, and prodded. Given that visual horror, a body count of "500" for American society was no longer the threshold.

Half a world away in a hotel room in San Francisco, President Clinton watched the images, horrified and angered, demanding to know who made the decision to undertake the mission, and why he had not been informed.

In Mogadishu, members of the task force gathered by a set, watching the morbid spectacle. Delta and Rangers waited for the word to launch again to secure the bodies. They now knew where the remains were. TF 160 pilots just wanted to go and mow down the cheering crowds. As tempting as it was, the commanders feared the cost in terms of local lives and nixed the idea.

Later that evening, American helicopters began to fly over Mogadishu, broadcasting messages to their captured comrade whom they knew was still held a prisoner somewhere below.

"Mike Durant, we will not leave you."

"Mike Durant, we are with you always."

"Do not think we have left you, Mike."

Hearing the messages, Durant felt some encouragement.

The following day, 5 October, A Company, 3rd Ranger Battalion, arrived in Mogadishu to augment the battered task force, remaining until 23 October.

The former U.S. Ambassador to Somalia under President George Bush, Robert Oakley, was dispatched to deliver the news: all efforts to capture Aidid would be called off and Task Force Ranger would be reinforced to stay on as a show of military resolve but would be withdrawn by March 1994. By Oakley's visit, all the badly mutilated bodies of those killed at the Super 64 site had been turned over. Only one major issue remained—the release of Durant, alive.

In that Aidid was still in hiding, Oakley met with the warlord's clan on 8 October and delivered President Clinton's message. When informed that the President wanted Durant released without conditions, the warlords informed Oakley that there would need to be a trade for the sixty to seventy clansmen that Task Force Ranger had already captured. Oakley offered to relay their request to the president for his consideration, but there would be no guarantee.

Durant needed to be released soon and, to assist the clan with its decision, the former ambassador offered a bit of friendly advice. He explained: When the decision is made to launch a rescue mission for the missing pilot, "the minute the guns start again, all restraint on the U.S. side goes. Once the fighting starts, all this pent-up anger is going to be released. This whole part of the city will be destroyed, men, women, children, camels, cats, dogs, goats, donkeys, everything."

The message and advice were delivered to Aidid, who quickly turned over the pilot—along with a Nigerian POW captured during Aidid's 5 September attack—on 14 October.

OBSERVATION

Despite the humiliation of having dead American soldiers publicly desecrated, despite evidence that Aidid had been struck a mortal blow, despite reports that Aidid's supporters were fleeing the city, despite knowledge that the Somali arsenal of RPGs was at or near complete depletion following the huge expenditure of rounds on 3-4 October, and despite overtures that other influential Somalis were offering to dump Aidid as a peace offer, the president and his staff lacked the guts to pursue anything further in Somalia. For then President William Jefferson Clinton, eighteen dead American soldiers was eighteen too many.

The United States, the world's one true superpower, had quit, withdrawing with its head down and tail between its legs. The courageous sacrifice made by the members of Task Force Ranger was for a cause totally unimportant to their commander-in-chief. Not only had their mission been called off the very next day, just as it was within their grasp to achieve success, but all of their previous efforts also went for naught. The president accepted Oakley's recommendation in support of the clan's request and ordered the release several weeks later of all those detained after their capture by Task Force Ranger. In hindsight, the eighteen American dead suffered by and in support of Task Force Ranger had turned out to be an exceptionally huge and insurmountable number.

With its lack of national resolve, the United States government actually encouraged extremist groups, such as Osama bin Laden's al Qaeda (al Qaeda is Arabic for "The Base"), to attack U.S. interests and its allies in the years to follow. Believing that the United States lacked the resolve to confront them, these terrorist groups launched a series of attacks that killed many more than would have been killed had the United States stayed the path in Somalia, not because the conflict in Somalia was resolvable—it wasn't—but because the strategic message to extremists was paramount—don't test the United States. Whether one agrees or disagrees with the U.S.'s commitment, once committed, we needed to stay the course.

In the days following the battle, a letter hand-written by the

Task Force Ranger commander, Major General William F. Garrison, was delivered to the president and secretary of defense through the House National Security Committee. The letter was composed of thirteen brief paragraphs in which he reported about the mission and its results. Paragraph 1 read: "The authority, responsibility, and accountability for the Op[eration] rests here in Mog[adishu] with the TF Ranger commander, not in Washington." Up front, the general had displayed that one aspect of character that eludes many national leaders today—personal responsibility. The buck stopped with him.

The best of America's fighting elite—Delta, SEAL, Para Jumpers and Ranger—had performed incredibly well in the face of overwhelming adversity. But even the Rangers had to admit that the older, more experienced "D-boy" Delta operators—many of whom were Ranger School graduates, were, with few exceptions, virtual killing machines.

Within the Ranger ranks, by a large margin the majority of the force, all performed their duties, but some performed better than others. Understanding that most of the junior enlisted Rangers were not Ranger School graduates and ranged in age from seventeen to nineteen, it should not be surprising that some found it difficult to face up to such adverse, self-testing, and life-altering circumstances. The fact that some men did not fight with total bravery does not denigrate them, nor does it make them cowards. In the words of one participant, "All gave some, but some gave all." Somalia showed that high intensity conflict does not have a monopoly on intense combat.

Nor does the questioning of their individual or group conduct indicate criticism. Whatever mistakes were made, those who were not there should check their impulse to see errors; it is most unlikely that they or anyone else could have, or would have performed better. However, despite the Ranger's elite status—or more appropriately maybe because of that elite status—there are lessons to be learned; and questions persist that deserve answers to provide these special forces every opportunity and advantage for future engagements.

Much has been made about this "failed" raid that did achieve the stated mission objective of capturing the two Aidid lieutenants. Unfortunately, there were flaws at the strategic, operational, and tactical levels that contributed to the final casualty figures. The raid certainly failed at the strategic level for the strategic objective as it should have been defined by national policy was never clearly stated. Deploying elite combat forces such as the Rangers and Detachment-Delta to conduct a citywide manhunt is not exactly one of their mission essential or war-fighting tasks—in other words it's not one of their trained specialties. Intelligence abysmally underestimated the tenacity of the enemy, the status of the warrior image within the Somali society, and the pride associated with that warrior heritage. Worse yet, was the failure to realize that the U.N. focus on the SNA and the warrant to arrest Aidid were, in the minds of Somalis, de facto declarations of war that forced their hands.

Operationally, there were a number of very critical errors with one of the most egregious being the lack of a designated chain of command that included both Major Generals Montgomery and Garrison. Unity of command would have brought the Quick Reaction Force (QRF) into the Task Force Ranger planning phase, thus preventing the waste of a considerable amount of time by the QRF while Task Force Ranger lay under siege in the streets. For example, QRF planning and preparing for rescue operations: these should have been ready to execute, but no one had a clue about what was involved. This lack of integration also seriously impacted Task Force David's ability to coordinate close-air support with TF 160 aircraft during the QRF's first deployment attempt, thus prompting the rescue force's withdrawal back to the American compound, causing further delay.

And, while some may raise questions about the unity of command at the tactical level on the ground, that seems to be less of an issue than claimed. Though there were three independent company grade commanders on the ground—Captains Miller, Steele, and the CSAR captain—the issue of "command" was, in the end, a non-issue. The rationale was quite simple: all were in touch with the senior ground commander at all times—first Lieutenant Colonel McKnight, then Lieutenant Colonel Harrell. A lieutenant colonel trumps a captain every time. If there were command issues at the local tactical level, they were a result of a failure to communicate or for the senior commander to make a command decision. It was not the result of a lack of clarity regarding the chain of command.

In addition to command and control operational breakdowns, there was another critical operational failure and that was in the realm of intelligence. The failure to identify the stockpiling of RPGs by the Somali militia was the intelligence community's first mistake. But, to overlook the realistic possibility that such weapons could be used to shoot down a helicopter—such as had occurred to a 10th Mountain Black Hawk late in September—ultimately proved to be a fatal, and deadly, interlaced oversight.

The deadliest error, though, proved to be the lack of armor or Spectre gunship support. There are four components to combat power: firepower, maneuver, protection, and leadership. Unfortunately for Task Force Ranger, leadership was about the only thing they had available to them for contingency operations. Having been denied armor support by Washington, the Quick Reaction Force had to rely initially on its own wheeled, soft-skinned vehicles to secure the downed aircraft. As events so clearly proved, American combat power was not up to the task. Consequently, American soldiers lost their lives or were wounded as the U.S. command sunk to the level of begging combat power from Malaysian and Pakistani forces to augment the Quick Reaction Force. Some may claim this is hindsight. Others would refer to it as a prudent measure to protect the force.

Though the task force's commander, Major General Garrison, testified before the Senate Armed Forces Committee that it was unlikely that armor and AC-130H Spectre gunships would have saved lives, an exceptionally strong and objective argument can be made to the contrary. Four of these gunships performed superbly during the period 9 June to 12 July when dealing with previous confrontational events. Why then, if these heavy weapon's platforms were stationed nearby in Djibouti, only 680 miles away, and still under Garrison's operational control, were they not brought in for support that fateful evening when Task Force Ranger needed every bit of suppressive fire it could muster? Within a few hours of flight, they could have been on the scene.

––––––––––

As for armor, let us compare the rebellion in Chechnya against the Russians and the task at hand in Mogadishu. To claim at that time that Mogadishu was another Chechnya is an invalid comparison for the Chechen rebels were much better trained and equipped than the Somali militia. Armed with heavy antitank weapons and powerful antitank mines—in addition to being well trained and well led—the former Soviet Republic rebels were able to make short work of the ill-trained, ill-equipped, and ill-led third-rate Russian units sent to subdue them.

Urban combat offers the opportunity to implement three different attack profiles against the weakest parts of an armored vehicle: side, rear, and top. As a consequence, tactical doctrine calls for dismounts to lead in a city environment with armor providing covering and supporting fires. The Russians failed to execute to standard. Furthermore, old-style Soviet BMPs or BRDMs—two different versions of Armored Personnel Carriers—were not designed to defeat RPG rounds. Poorly maintained and obsolete T-55 tanks with fatigued armor that had grown more brittle with age and limited dismounted infantry support were able to be defeated by either heavy weapons, command-detonated or tilt-rod mines (mines that have a two-foot rod extending straight up that does not require a vehicle to actually run over it with a wheel or track to detonate them), or multiple RPG shots directed against the weakest part of the vehicle.

Those who profess that armor would have made no difference based on the Chechen model should not confuse American military might against a poorly trained militia armed with low-tech weaponry with the results of a tired and worn out Russian Bear against a very capable, heavily armed, relatively well-trained, and highly disciplined paramilitary organization. The U.S. M2 Bradley Infantry Fighting Vehicle (IFV) is specifically designed to deflect or prevent RPG rounds from penetrating its sloped armor, and armored tread skirts, as demonstrated during live-fire demonstrations during the IFV's development. Its 25-mm chain gun—a small caliber cannon that can fire nearly as quickly as a machine gun because of its linked ammunition—with armor piercing rounds and M-240G 7.62-mm coaxial gun are mounted in a 360-degree revolving turret that can be fired in a nearly vertical position—a significant advantage in urban combat. An M-2 .50-caliber machine gun and three firing ports on each side of the vehicle to house shortened versions of the M-16 provide additional offensive and defensive firepower. The Tube-launched, Optically-tracked, Wire-guided (TOW) anti-tank missile launcher is just icing on the cake to take out any fortified targets. Combined with its optics for long-range viewing and night-seeing capabilities, the vehicle makes for the best mechanized urban fighting vehicle in the world.

As for the combat-loaded, seventy-ton M1A2 Abrams main battle tank, its unique Chobham armor (named after the location in England where the laminated armor was first created) is designed to defeat any round fired with the exception of the latest Russian main battle tank or another Abrams—as was unfortunately demonstrated in the first Gulf War. Armor skirts that cover much of the tank's treads make it nearly impossible to immobilize the tank. An RPG probably wouldn't even put a dent in the behemoth much less stop or kill it. As for firepower, a 120-mm cannon, a M-240G 7.62-mm coaxial machine gun, a turret-mounted 7.62-mm M-60 machine gun, and an M-2 .50-caliber machine gun would make it the biggest gun on the street. And, if there is anything the militia in Somalia was impressed by and respected, it was big guns—the bigger the gun, the greater the respect and subsequent apprehension. This is basic battlefield psychology.

The combined capabilities of infantry and armor within an urban setting were eventually amply demonstrated during Operation Iraqi Freedom (OIF) on 7 April 2003. The 2nd Brigade Combat Team (BCT) of the 3rd Infantry Division (Mechanized) had advanced two task force battalion-sized elements into the very heart of Baghdad.

Highway 8, a four lane road leading in from the south, was the main supply route for the BCT and needed to be kept open against a determined foe who was trying to cut off the two BCT task forces within the city. That task fell to the BCT's final task force, the 3rd Battalion, 15th Infantry (TF 3-15). Comprised of two mechanized infantry companies of M2 Bradleys, a tank company of M1 Abrams and elements of a combat engineer battalion, TF 3-15 advanced into the heart of Baghdad to seize, secure and defend three key intersections on the highway that they named "Moe, Larry and Curly"—after the Three Stooges.

The section of Highway 8 leading to Moe, Larry and Curly had come to be known as "RPG Alley" and the area was defended by a combination of Saddam Hussein loyalists—Special Republican Guards and Fedayeen militiamen—and suicidal Syrians intent on fighting a Jihad (Holy War) against the American infidel. Though the attacks were uncoordinated and lacked a central command, the enemy's tactics seemed to have taken a page out of the 1993 Somali success against Task Force Ranger in Mogadishu. Clad mostly in civilian clothes, firing from residential buildings and mosques, the Iraqi and Syrian fighters launched a deluge of RPGs and suicide bombers in bomb-laden civilian cars at the embattled Americans in an attempt to cut them off and isolate them on city streets as they poured in reinforcements from trench systems carved into the rubble and blasted from surrounding buildings, parks, and intersections in an effort to inflict maximum casualties.

The battle waged around the three objectives for nearly ten hours as wave after wave of suicidal enemy soldiers driving civilian cars, trucks and buses at speeds up to 70 mph attempted to crash against the American armor. Tanks and Bradleys ran low—and some even ran out of—ammunition. Flame and smoke obscured the field of battle allowing the irregulars to get within hand grenade distance of the Americans. On Objective Moe, every vehicle was hit by at least one RPG—many three or four times. Only seven U.S. soldiers were wounded at that objective with one of them wounded seriously enough to be evacuated. On Objective Larry, the company commander watched an Abrams and two Bradleys be struck by RPGs with minimum damage. His own tank was struck five times by RPGs and still remained operational.

The attacking force was estimated to be at least 900 in number. Of that total, 350 to 500 were killed and another fifty taken prisoner. Despite the fierceness and long duration of battle, however, only two U.S. soldiers manning machine gun turrets in a thin-skinned supply convoy were killed in action by RPG rounds; thirty soldiers were wounded.

Though the tactics used by the enemy in Mogadishu and Baghdad were strikingly similar, the results were strikingly disparate. This time, the U.S. troops in Baghdad had armor, which was more than a match for the ever-present enemy weapon of choice, the rocket-propelled grenade.

The task force's senior NCO, Sergeant Major Robert "Black Hawk" Gallagher, was a special forces veteran who'd been wounded three times during the battle in Mogadishu nearly a decade before. Gallagher observed that:

> In both places, U.S. soldiers faced persistent irregulars who used buildings as cover, and the severity of the [gunfire] was the same. In Baghdad, however, the big difference was that we had armor. That reduced the amount of casualties we had significantly.

The lessons learned later in Baghdad aside, the respect that the Somalis had for armor had already been demonstrated prior to October 1993 during an incident on 17 June, when a Somali mob demonstration that day—a harbinger of events to occur on 3 October—was stabilized by the show of force exhibited by light tank and Armored Personnel Carrier (APC) reinforcements. One can only begin to imagine the impact a behemoth American Abrams main battle tank would have had.

Furthermore, one can only conclude that American armor would certainly have made a huge difference, preventing or diminishing the loss of those eighteen KIA, and one does not have to look any further than that mission, itself. The proof was demonstrated on the streets of Mogadishu that horrendous day when the lightly armored convoys of McKnight, Struecker, and David managed to return with all or most of their thin-skinned vehicles—despite the overwhelming number of RPGs fired at them. Kallman survived a direct hit on his Humvee, yet he drove on. An RPG blast beneath one of Struecker's Humvees at the K-4 circle did not knock it out. That's confirmation.

Even the old BRDMs (Soviet style APCs) in the multinational rescue effort survived RPG hits, with the exception of the lead two vehicles which took wrong turns and found themselves isolated and vulnerable to close assault. As for the M-48s, the only force stopping these tanks was the timidity of the drivers to breach the roadblocks. RPGs were no problem.

Therefore, considering the survivability of much less capable vehicles on the hostile streets of Mogadishu, it is baffling to think that an American-trained platoon of Abrams with a company of Bradleys would not have made a difference. Such equipped units are pure combat power—firepower, maneuver, and protection—especially in the harsh urban environment of Mogadishu against an undisciplined para-military rabble.

Moreover, and this cannot be overemphasized, the psychological impact of armored vehicles is significant. They are intimidating machines and surely have an impact in cowing any opponent—professional or irregular. They would have been unstoppable, crashing through any barrier thrown before them, as the significantly less capable Humvees were able to do. They would have been responsive and easily employed to travel and fight at night, if necessary. Should there have been a fortuitous hit that immobilized a vehicle, it would have been a simple process to hook the disabled vehicle up to an operable Abrams or Bradley with a tow bar—which could have been anticipated and in place for the operation—to drag the immobile vehicle along. American armor would most probably have been able to get to both crash sites in time to save the lives and limbs of many of the trapped warriors, had they been stationed only two to three miles away as part of the QRF.

The finest armor in this particular urban environment would have been an invaluable resource. Unfortunately, the main problem with many officers schooled for so long primarily in light infantry tactics is that they are not fully knowledgeable of armor capabilities. While armor has some obvious disadvantages primarily associated with its immense size and weight, armor actually provides a powerful and exceptionally capable urban combat force when used properly with dismounted infantry. The successful use of armor when protected by infantry in urban combat was proven and well documented as early as World War II by the Germans, British, Americans, and the Soviets. It was a lesson that we seemed to have forgotten.

A complete analysis of the operation also indicates some tactical flaws. For sheer simplicity, a review of Rogers' Rangers Standing Orders shows a violation of three of the nineteen principles:

- **Rule 1:** *Don't forget nothing:* The force left Night Observation Devices (NODs) and water behind, believing the raid would be quick. Yet it was past 3:30 PM when the raid was launched—not much daylight was left should anything go wrong.

- **Rule 4:** *Tell the truth about what you see and what you do—don't ever lie to a Ranger or officer:* It has been reported that Staff Sergeant Eversmann, leader of Chalk 4, was unable to obey an order to move, though he acknowledged that he would do so. Only by luck and accident did Lieutenant Colonel McKnight discover Eversmann and his men as he drove by the squad's position with his convoy. If it weren't for the fortuitous moment, there would most likely have been an additional twelve Rangers added to the KIA total.

- **Rule 11:** *Don't ever march home the same way:* The intent behind this rule is to ensure one takes a different route to avoid a possible ambush. It can also be interpreted as: never conduct an operation the same way twice. Some different techniques that could have been employed to change the pattern: false insertions into other target areas to create the illusion of multiple targets; deception plans to confuse the enemy into believing another objective was the primary target of interest; launch an actual supporting raid on another objective prior to the main effort raid to disorient the enemy; minimize the employment of helicopter assets where they are not as critical against outlying targets (such as the previous six raids) to minimize the enemy's focus on trying to create a means to shoot them down. The real issue was that Task Force Ranger had not planned for the unexpected, and the unexpected in this instance was that it would not capture the Somali Warlord Mohamed Farrah Aidid in the first few raids. By exposing all its capabilities up front with the initial raids, it had nothing left to surprise the Somalis with when it really counted. It was the equivalent of a dealer showing all of his cards before his opponent drew his.

There were also serious communications shortcomings. The inability of the Delta and Ranger units to communicate effortlessly, and the use of un-enciphered messages back and forth was, in fact, a blunder of the first order. For the strongest and most technologically advanced military force in the world, this seems inexcusable. Surprise, stealth, speed, and flexibility are paramount to the survival of these elite and lightly armed special forces. Such units cannot afford time nor can they afford confusion and yet, because of the abysmal communications systems and processes, they lost valuable time and suffered from mass confusion. Considering the advanced technology available today—already reflected in the communications gear carried by each Delta operative—the financial and material resources available to the United States and the criticality of the missions assigned the Ranger Regiment, each Ranger going into a mission should be provided with the ultimate in fighting gear. The advantages in terms of the military accomplishment of national objectives more than compensates for the financial costs associated with such equipment. What's an American warrior's life worth?

It is said that Japan had "Victory Disease," thinking it had executed a successful attack on Pearl Harbor—only to find that it had "awakened a sleeping giant" that ultimately left the Japanese homeland in ruin and shrouded two major cities in radioactive haze. The Israelis would later formulate "The Concept," believing that no Arab nation, or even group of Arab nations, would have the audacity to attack them following their stunning victory in the 1967 "Six-Day War"—only to have the 1973 "War of Atonement" shatter that misplaced belief—and almost cause the downfall of the State of Israel.

Success and victory sometimes breed complacency, and America apparently had a touch of that in Somalia based on our successes in Grenada, Panama, Operation Desert Storm, and the Iraqi Kurdish Operation Provide Comfort. This "disease"—or even hubris, false pride—has a way of getting those so infected into trouble, no matter how great they may think they are.

This hubris led to the greatest Task Force Ranger failure of all—it failed to expect, plan, and train for the unexpected. It failed to plan for what would happen if it didn't quickly capture Aidid. It failed to plan and train for multiple aircraft being shot down. It failed to plan and train the QRF response. United States commanders in Mogadishu had the Victory Disease on 3 October 1993 and paid for it—or rather, eighteen of their finest spilled their life's blood to pay for it.

Amazingly though, despite the communication gaffes, despite the misapplication of military force and lack of relevant resources, it almost worked! America was close to meeting its objective, if that objective was to remove Aidid from the political and military equation. Task Force Ranger had Aidid on the ropes. Unfortunately, just at the moment that the United States needed a show of intestinal fortitude, the president of the United States cut and ran, thus making the personal and professional sacrifices of all worthless.

Sadly, the U.N. ambassador to Somalia, U.S. Admiral (Retired) Jonathan Howe, would later admit Task Force Ranger soldiers "had many opportunities to eliminate [Aidid]. That's not our job. We're trying to arrest him." Maybe, based on the outcome of our experience in Somalia, some thought it should be taken in the future to make eliminating such an individual "our job." At times, it's best to make it personal, if you are looking to get someone's attention.

Setting aside major interrogatories, as a country what has been relatively ignored, unfortunately, in the national debate is the heroism and valor of these soldiers. Sadly lost is the personal sacrifice of these men. Members of the profession of arms do not ask to be placed in harm's way. They do not ask to be placed in circumstances that find them significantly outnumbered, outgunned, and deep within enemy territory. They do not ask to be placed in situations that require them to suffer wounds and death in defense of their comrades.

While they do not ask for any of these hardships, soldiers freely endure them anyway in a demonstration of incredible bravery and devout professionalism. They endure because of dedication. They endure because of responsibility. They endure because it is their duty. They endure because they believe in the Creed. They endure because many wear the coveted black and gold Ranger Tab on their left shoulder.

For their actions that day the two Ranger-qualified Delta snipers, Master Sergeant Gary I. Gordon and Sergeant First Class Randall D. Shughart, were posthumously awarded the Medal of Honor in a White House ceremony on the morning of 23 May 1994. It was reported that Sergeant First Class Shughart's father, Herb Shughart, refused to shake the president's hand. Later that day, the two Rangers were enshrined in the nation's Hall of Heroes at the Pentagon.

Regrettably, having to compete with the burial of Mrs. Jackie Kennedy Onasis, the few square inches of newsprint and the minute or so of television sound-bite dedicated to their monumental sacrifice were a great disservice to them and their families. These American heroes were faced with the choice of death or dishonor—as they saw it. Their standards of dedication, responsibility, and duty left them with no other option but to knowingly choose to sacrifice themselves.

That call to honor should not be so easily dismissed, nor ever forgotten. Facing overwhelming forces deep within enemy territory, the ground and air members of Task Force Ranger more than held their own against such incredible odds. "I will never leave a fallen comrade to fall into the hands of the enemy." The Rangers and Delta forces never agonized over their decision to remain with the body of the dead pilot. Though the Ranger forces could have fought their way out to a more secure site, bringing their wounded with them, that alternative would have meant leaving their dead comrades behind for Aidid's jackals to tear apart.

In May of 1941, the staff of British Admiral Sir Andrew Browne Cunningham recommended that the Royal Navy abandon British ground forces on Crete in order to protect the Royal Navy's ships. The admiral resoundingly dismissed its recommendation: *It takes the Navy three years to build a ship. It would take three hundred to rebuild a tradition.*

Rangers live by their own tradition, their own ethic, their own Creed. Though some may question their decision to remain at the crash site, others understand that the Ranger Creed is a function of tradition and sacrifice; it is not a calculus. True to their Creed, the members of Task Force Ranger who secured the wreckage of Super 61 remained at the crash site to retrieve the body of a fallen comrade while Master Sergeant Gordon and Sergeant First Class Shughart most likely knowingly and willingly sacrificed their lives in the defense of Super 64. There are those who might question the judgment of risking lives on the behalf of those who have already given theirs. They cannot understand the pride and loyalty intrinsic with being a Ranger and probably they never will. So be it.

For an article published in the August 1, 1994, edition of *U.S. News & World Report*, Master Sergeant Gordon's widow was asked to write a letter to her children on a topic she believed to be important to their development. Mrs. Carmen Gordon elected to write to her two children about "Responsibility" and what it meant to their father.

> I hope that in the final moments of your father's life, his last thoughts were not of us. As he lay dying, I wanted him to think only of his mission to which he pledged himself. . . . He chose the military, and "I shall not fail those with whom I serve" became his simple religion.

No man or woman can read that piece and not feel a tugging of the heart, a swelling in the eyes, and a deeper appreciation for the term "responsibility." Master Sergeant Gordon's widow is a true soldier's wife who understands what it means to be a Ranger.

Beyond the tragedy of the lives lost, bodies maimed, and families devastated, there is a reverberating legacy of Mogadishu that resonated on a national—and even global—scale for nearly a decade. It was not until the launch of the Global War On Terrorism (GWOT) following the events of 11 September 2001 that the United States understood and began to implement the lessons learned that have been bought and paid for by the blood of valiant American warriors. This shameful legacy was exhibited during an interview by ABC News on 28 May 1998 of Osama bin Laden, who noted:

> We have seen in the last decade the decline of the American government and the weakness of the American soldier who is ready to wage Cold Wars and unprepared to fight long wars. This was proven in Beirut when the Marines fled after two explosions. It also proves they can run in less than 24 hours, and this was also repeated in Somalia.

Bin Laden's citing of Somalia was premeditated, relevant, and quite revealing. Not only did it serve as an example of the terrorist's disdain for, and derision of America's national will, it also provided the rationale for the terrorist bombings of the Khobar Towers in 1995, the destruction of the U.S. Embassies in Tanzania and Kenya in 1998, and the attack on the USS Cole in 2001. All these terrorist acts hearken back to the U.S. decision to pull out of Somalia after suffering a relative handful of casualties, just as victory was in its grasp. Given that cowardly decision, it took little to convince non-conventional adversaries that an exceptionally productive way to drive out the United States or to gain an advantage is wantonly to kill Americans, including many innocent local bystanders. For terrorists, life is very cheap.

The American resolve in world opinion was, at this time, in question as a result of the actions taken following that fateful day in Mogadishu. Also in question were the attitudes of our national and military leaders, as well as the effectiveness of American troop deployments. Since 3 October 1993, "force protection" had served as the byword of all American leaders, for all deployments. Unfortunately, the current definition of "force protection" has taken on a distorted, and nearly perverted perspective that has become synonymous with playing it safe.

Do our officers, military planners, and political leaders need to worry that if there is a casualty their careers are over? Is it all right for soldiers to be loaded down in restrictive gear—Kevlar helmets and flak vests—while fearful leadership minimizes their contact and exposure to ensure that no one gets hurt? Will such mindsets defeat the Bin Ladens of this world who can inspire men—and women—to commit suicide in the name of Islam, or will it lead United States forces into long-term occupations, enforcing false, constricted borders, all of which are not in the national interest?

Even with the media exposure provided during Operation Enduring Freedom (OEF) in 2001 and Operation Iraqi Freedom (OIF) in 2003, the American public does not grasp the reality of modern combat. It has been enamored by the overwhelming success of the United States Armed Forces since the Gulf War, captivated by the video game approach to combat that high-tech weaponry demonstrates, and given a false expectation that all combat is sterile and impersonal. Modern conflict is not all overwhelmingly successful, nor is it all high-tech and impersonal. Modern conflict is overwhelmingly brutal, in-your-face, and very personal. And, the American public should learn, and should realize, and should accept, that with real war comes bloody noses. There will be, to paraphrase Colin Powell, the potential for a few "bad days." Force protection is force projection, and as the world's superpower, the United States, should begin to act as a military superpower. To protect by force of arms requires always the guts to accept losses by projection. That's a class act befitting a great nation.

American society will generally accept that United States soldiers are placed in harm's way and even killed to pursue national policy when the good fight becomes necessary. Was there, in Somalia, a good reason to fight? There were no vital national interests at stake, and U.S. security was not an issue. In hindsight, it would seem that America's involvement was purely an altruistic effort to bring order and stability to a region that was suffering the loss of tens of thousands of innocent victims in a brutal civil war. From a purely moral and ethical point of view, that could be a good reason to fight. Is this an end that can justify the means?

But there was just one, obtrusive, trifling concern never fully addressed by America's leaders nor questioned by the media. Many of those apparent innocent "victims" did not care for U.S. or U.N. help. Many of those victims, in fact, were more inclined to spit upon the very forces attempting to protect and assist them—a display of scorn that a large number of the populace routinely carried out whenever U.S. or U.N. forces were about.

Though, as of 2004, they were finally attempting to make what seems to be an honest effort to snuff out the civil war, Somalia was, and still is a void in Africa, not a nation. Its society, if one can call it that, can feel little for the greater good given how its daily rituals focus on nothing beyond the family unit, food, and basic survival. Building a country from the outside is doable; creating a nation, however, must come from within. True peacemaking and nation building require more of the beneficiaries than of the benefactors. More importantly, the beneficiaries must want the same kind of assistance that the benefactor is offering to provide. That was never the case in Somalia in 1993 and little has changed since then.

Why, then, is it deemed incumbent on America to step in and sacrifice the blood and lives of its warriors in a philanthropic attempt to provide assistance to beneficiaries who lack the will, ability, and desire to sustain such efforts when there is no vital United States interest at stake? Ultimately, the greatest lesson learned from Somalia needs to be the realization and acceptance of the fact that when a nation's people do not give a damn about their country or their own quality of life, neither should the people of the United States.

History is replete with examples of defeated armies that attempted to force those who believe power is derived from the barrel of a gun into giving up those ideals in pursuit of some abstract concept titled nation-state legitimacy. It didn't work for empires in centuries past and, as demonstrated on 3 October 1993, it didn't work for the U.N. or the U.S. either. The warlords and their ilk still live on in Somalia.

Master Sergeant Gary I. Gordon

MEDAL OF HONOR CITATION

GORDON, GARY I. (Posthumous)

Rank and organization: *Master Sergeant, U.S. Army.*

Place and date: *October 1993, Mogadishu, Somalia.*

Born: *Lincoln, Maine.*

Citation

Master Sergeant Gordon, United States Army, distinguished himself by actions above and beyond the call of duty on 3 October 1993, while serving as Sniper Team Leader, United States Army Special Operations Command with Task Force Ranger in Mogadishu Somalia. Master Sergeant Gordon's sniper team provided precision fires from the lead helicopter during an assault and at two helicopter crash sites, while subjected to intense automatic weapons and rocket propelled grenade fires. When Master Sergeant Gordon learned that ground forces were not immediately available to secure the second crash site, he and another sniper unhesitatingly volunteered to be inserted to protect the four critically wounded personnel, despite being well aware of the growing number of enemy personnel closing in on the site. After his third request to be inserted, Master Sergeant Gordon received permission to perform his volunteer mission. When debris and enemy ground fires at the site caused them to abort the first attempt, Master Sergeant Gordon was inserted one hundred meters south of the crash site. Equipped with only his sniper rifle and a pistol, Master Sergeant Gordon and his fellow sniper, while under intense small arms fire from the enemy, fought his way through a dense maze of shanties and shacks to reach the critically injured crew members. Master Sergeant Gordon immediately pulled the pilot and the other crew members from the aircraft, establishing a perimeter which placed him and his fellow sniper in the most vulnerable position. Master Sergeant Gordon used his long range rifle and side arm to kill an undetermined number of attackers until he depleted his ammunition. Master Sergeant Gordon then went back to the wreckage, recovering some of the crew's weapons and ammunition. Despite the fact that he was critically low on ammunition, he provided some of it to the dazed pilot and then radioed for help. Master Sergeant Gordon continued to travel the perimeter, protecting the downed crew. After his team member was fatally wounded and his own rifle ammunition exhausted, Master Sergeant Gordon returned to the wreckage, recovering a rifle with the last five rounds of ammunition and gave it to the pilot with the words, "good luck." Then, armed only with his pistol, Master Sergeant Gordon continued to fight until he was fatally wounded. His actions saved the pilot's life. Master Sergeant Gordon's extraordinary heroism and devotion to duty were in keeping with the highest standards of military service and reflect great credit upon him, his unit, and the United States Army.

Sergeant First Class Randall D. Shughart

MEDAL OF HONOR CITATION

SHUGHART, RANDALL D. (Posthumous)

Rank and organization: *Sergeant First Class, U.S. Army.*

Place and date: *October 1993, Mogadishu, Somalia.*

Born: *Newville, Pennsylvania.*

Citation

Sergeant First Class Shughart, United States Army, distinguished himself by actions above and beyond the call of duty on 3 October 1993, while serving as a Sniper Team Member, United States Army Special Operations Command with Task Force Ranger in Mogadishu, Somalia. Sergeant First Class Shughart provided precision sniper fires from the lead helicopter during an assault on a building and at two helicopter crash sites, while subjected to intense automatic weapons and rocket propelled grenade fires. While providing critical suppressive fires at the second crash site, Sergeant First Class Shughart and his team leader learned that ground forces were not immediately available to secure the site. Sergeant First Class Shughart and his team leader unhesitatingly volunteered to be inserted to protect the four critically wounded personnel, despite being well aware of the growing number of enemy personnel closing in on the site. After their third request to be inserted, Sergeant First Class Shughart and his team leader received permission to perform this volunteer mission. When debris and enemy ground fires at the site caused them to abort the first attempt, Sergeant First Class Shughart and his team leader were inserted one hundred meters south of the crash site. Equipped with only his sniper rifle and a pistol, Sergeant First Class Shughart and his team leader, while under intense small arms fire from the enemy, fought their way through a dense maze of shanties and shacks to reach the critically injured crew members. Sergeant First Class Shughart pulled the pilot and the other crew members from the aircraft, establishing a perimeter which placed him and his fellow sniper in the most vulnerable position. Sergeant First Class Shughart used his long range rifle and side arm to kill an undetermined number of attackers while traveling the perimeter, protecting the downed crew. Sergeant First Class Shughart continued his protective fire until he depleted his ammunition and was fatally wounded. His actions saved the pilot's life. Sergeant First Class Shughart's extraordinary heroism and devotion to duty were in keeping with the highest standards of military service and reflect great credit upon him, his unit, and the United States Army.

A UH-60 Black Hawk high over the city of Mogadishu. (1993)

The devastation, destitution and turmoil of the Somali capital is evident with this overhead view of Mogadishu. (1993)

Rangers "Fast Rope" from UH-60 Black Hawks at a Military Operations in Urban Terrain (MOUT) training facility.

The only ground photograph declassified and released to date of the 3 October firefight in the streets of Mogadishu. (3 Oct. 1993)

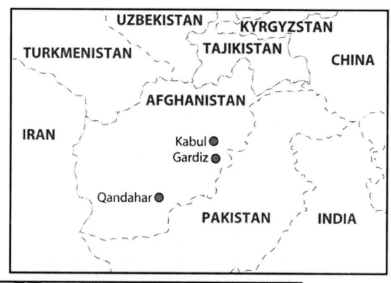

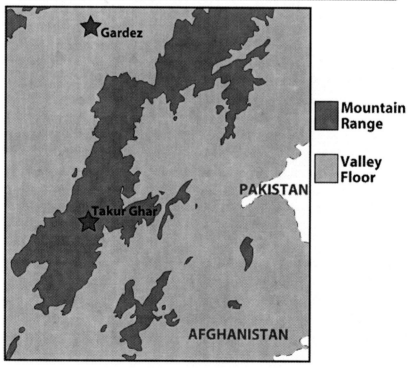

Eighteen

BROTHERHOOD

We few, we happy few, we band of brothers; For he today that sheds his blood with me, Shall be my brother.

—William Shakespeare
King Henry V, Act 4. Scene III
(English writer and playwright, 1564–1616)

DATE: *4 March 2002*

LOCATION: *Takur Ghar Mountain, Afghanistan (also referenced as "The Battle of Shah-I-Kot")*

MISSION: *Operation Enduring Freedom (OEF) - a twenty-three-member Army Ranger Quick Reaction Force (QRF) is deployed to rescue a Special Operations Forces (SOF) reconnaissance team trapped on a mountaintop under heavy enemy fire.*

BACKGROUND

FOR NEARLY TWO MONTHS, the United States had monitored major activity of al Qaeda and Taliban forces in the Shah-e-Kot valley, just southeast of Gardez, Afghanistan. To eliminate this enemy concentration, an operation code-named "Anaconda" was planned. In support of Anaconda, several Special Operations Forces (SOF) reconnaissance teams were inserted to establish observation posts (OPs) at key tactical locations along a mountain range to observe and direct air strikes against enemy forces. One such team was inserted near Helicopter Landing Zone (HLZ) Ginger, a point just below the Takur

Ghar Mountain ridge from which it could observe the southern tip of the valley.

A dominating piece of terrain, Takur Ghar rose to a height of 10,200 feet. From its peak, a 360-degree panorama provided an unobstructed view on clear days for fifteen miles. The ridge was sporadically covered with foliage, and the ground was white, covered by three feet of snow. To the mountain's east and west, there were almost sheer drop-offs. Despite the mountain's size, however, there was only one small landing zone to offload a helicopter within thousands of feet of the ridgeline.

The plan called for two reconnaissance teams to be inserted the night of 3 March by two MH-47E Chinook helicopters from the 2nd Battalion, 160th Special Operations Aviation Regiment (SOAR)—the "Night Stalkers" . . . also referred to as Task Force 160 or TF 160. The MH-47E code named "Razor 04" would insert its team to the north, while the other Chinook, "Razor 03," would insert its team on Takur Ghar.

Operational and maintenance problems delayed the mission, forcing the Razor 03 SOF team of seven—six SEALs (Navy Sea, Air, and Land) and one Air Force Special Operations combat control team member (CCT)—to be flown directly to the Takur Ghar summit in the early hours of 4 March to enhance speed at the cost of a stealthier ground approach from another point of insertion. This risk seemed acceptable at the time, for Takur Ghar was thought to be uninhabited.

THE LEGACY

At approximately 3 AM on 4 March, with tail ramp down, the twin-rotor black silhouette of the 52-foot long Razor 03 helicopter approached HLZ Ginger located in a small saddle just beneath the peak of Takur Ghar and touched down. Despite the darkness, both Chinook pilots and the team in the cargo bay could see fresh tracks in the snow. Goatskins lay about and other signs of recent human activity could be observed. The chopper's crew reported the presence of a heavy machine gun about 50 yards off the nose of the aircraft, but it was unmanned; such sights were not uncommon in Afghanistan, where caves and ridges were littered with seemingly abandoned tanks, armored personnel carriers, and anti-aircraft guns. On the ramp, Navy Petty Officer 1st Class Neil C. Roberts stood first in line to disembark. Despite the tell-tale signs of human activity, the SEALs announced that the mission was a go, and that they were departing.

Suddenly, the muzzle flashes of heavy machine-gun fire erupted from several directions, with bullets stitching the fuselage. Streaking from the left side, a rocket-propelled grenade (RPG) impacted against the side of the aircraft penetrating the skin, and detonating within the cargo bay in a blinding flash. Severed hydraulic and oil lines spewed slippery fluids along the floorboards and about the ramp area.

In the right rear of the aircraft, a crew chief yelled to the pilots, "We're taking fire! Go! Go! Go!" The helicopter shuddered and jerked as the pilots fought to lift the stricken chopper out of the hot landing zone (LZ) with its electrical system short-circuited and hydraulics seriously damaged. In a jerk, it began to lift off from the small clearing, gaining altitude.

In the rear, already off-balance from the impact of the RPG and losing traction from the fluids pouring on the floor about him, Roberts fell off the ramp with the sudden burst of engine power. A rear crew chief tried to grab the SEAL as he went off the ramp but went over the side himself as he lost his own balance and footing. Roberts fell approximately ten feet to the snowy ground below. The crew chief, tethered to the bird with a harness, dangled off the edge of the ramp. A second crew chief struggled to pull him back in as the helicopter was placed in a dive down the mountainside.

Leveling off, the pilot tried to climb, but the MH-47E began to shake badly from the damage caused by the bullet holes in its rotors and the hydraulics started to give out. Informed that Roberts had fallen out and been left back on top of Takur Ghar, the pilot attempted to turn back, only to find his controls locked up as the hydraulic fluid reservoirs ran dry.

With the dangling crew chief finally safely on board, the crew began pumping desperately to force spare hydraulic fluids back into the system. Though the controls came back and the pilot was able to level off the helicopter, the pilot knew that the chopper was no longer functional enough to return to combat.

"Sorry guys, we're going to have to abort," was all he could offer as they began to make their way farther down the mountain's slope.

Heavily damaged, Razor 03 limped north, eventually landing at the north end of the valley, approximately four miles from HLZ Ginger. Within 30 to 45 minutes, Razor 04 and its SEAL team arrived to help secure the downed aircraft. Options were discussed but with two teams, one operable aircraft, and the mistaken perception that enemy forces were quickly closing in on their site, the only viable option seemed to be to return with everyone to a staging base at Gardez, drop off Razor 03's crew, then return to Ginger with the SEALs for Roberts. The SEALs knew that time was of the essence and that their only real hope of getting Roberts out alive was to take their chances and be set down in the same LZ where Roberts had fallen.

It was about 5 AM when Razor 04 made its approach to Takur Ghar. Darkness was waning and about 40 feet above the HLZ the pilot saw the first muzzle flashes of a machine gun just off the nose of the helicopter. The MH-47E began to take heavier fires as it set down on the HLZ, the rounds "pinging and popping through." With machine gun fire cutting through the skin of the aircraft, the six member rescue team quickly exited the chopper. Though damaged, Razor 04 was able to lift off and safely return to base.

The SEALs moved out and took withering fire. All about them, the enemy was nowhere to be seen, located within dug trenches, hidden under trees, and obscured by shadows. They moved toward the most prominent features on the hill, a large rock, and tree at the base of which, unbeknown to them, were two enemy bunkers.

Approaching the tree, Technical Sergeant John Chapman, the team's CCT, and a nearby SEAL opened fire killing two of the enemy that the airman had spotted in a fortified position under the tree. From 20 yards away, another bunker position opened fire, catching Chapman, thirty-six, husband and father of two, with a burst of gunfire. The airman fell to the ground, mortally wounded.

In quick succession, two more SEALs were wounded by enemy fire and grenade fragmentation. Now, badly outnumbered by al Qaeda, with half the team either dead or wounded, and caught in a deadly crossfire on the mountain top, the SEALs decided to disengage from the direct firefight and began to withdraw down the peak.

Prior to Razor 04's insertion, the Ranger Quick Reaction Force (QRF)—from A Company, 1st Battalion, 75th Ranger Regiment—located at Bagram, approximately 100 miles from Gardez, had been alerted. The QRF was a designated standby unit trained for just such emergency rescue situations. Task organized into two elements, the QRF's Chalk 1—a military term and designation for a ground unit assigned to an aircraft—was assigned to the MH-47E code-named Razor 1, crewed by eight members of the 160th SOAR.

The overall commander, Captain Nathan Self, twenty-five, a West Point graduate, was assigned to Chalk 1. While they had been in Afghanistan since December and scrambled on a number of occasions, none of the team members had seen combat in Afghanistan, or anywhere else for that matter.

An additional eight Rangers and four Air Force special operatives from the Special Tactics Squadron completed Chalk 1. The aircrew of Razor 01 was composed of nine additional soldiers, a mixture of 160th SOAR and Air Force personnel. The remainder of the QRF, Razor 2's Chalk 2, was comprised of ten Rangers under the command of Staff Sergeant Arin Canon.

Shortly after 5 AM, as the SEALs—inserted by Razor 04—were fighting for their lives, Razor 01 and Razor 02, loaded with Chalk 1 and Chalk 2 respectively, lifted off from Bagram to make the flight to Gardez. Given the rapidly changing situation, the lack of intelligence gathering platforms in the area and the communications problems, the QRF flew off into the unknown, lacking any awareness of what was actually happening on Takur Ghar.

As the QRF flew to Gardez, the embattled SEALs requested the QRF's immediate assistance as they began to withdraw down Takur Ghar. Headquarters approved the request and directed the QRF to proceed quickly to the area and insert the team at an "offset" HLZ—meaning they were to select an LZ at their discretion but some distance away from the designated coordinates that had been provided earlier. They were not to land at HLZ Ginger, where Razors 03 and 04 had taken heavy fire and sustained serious damage.

Tragically, the QRF never received the "offset" instructions. Headquarters made several attempts to pass the information about the offset and SEAL situation through other means, but those efforts were misunderstood. At 5:45 AM, Razor 01 and Razor 02 were entering "the box"—operational area—of Anaconda. Razor 02 moved off to take a holding pattern over the Shahikot valley as Razor 01 flew on, directly towards HLZ Ginger. Soon, all communications were lost between the two MH-47Es as Razor 01 disappeared up, into the mountain.

Self and his men were totally uninformed of the situation. They did not know how many of the enemy to expect. They did not know what weapons systems they'd be facing. They didn't know exactly where the enemy would be. They didn't have any communications with support elements. And they didn't even know where the SEALs were that they were sent to rescue. They were essentially flying blind. "Into the valley of Death rode the 22" . . . to paraphrase Alfred Lord Tennyson's "The Charge of the Light Brigade."

The Rangers in Razor 01 were quickly losing the advantage of darkness as they continued to fly in the dawn's rapidly growing early light. The six-minute warning—a time hack reference from pilot to passengers—was issued to Chalk 1 at 6:09 AM. Approaching the HLZ from the south, the black form of Razor 01, illuminated by the sun, was just beginning to crest the mountains to the east. The helicopter was on its final approach when the impact of an RPG knocked out the right engine with the aircraft still 20 feet off the ground. The RPG was followed immediately by heavy machine gun fire that sprayed the side of the Razor 01 and shattered the glass in the cockpit. Small arms fire struck from three sides. Sergeant Philip J. Svitak, 31, father of two, flight engineer and right forward gunner fired a single burst of his 7.62-mm minigun before he was struck down by an AK-47 bullet and died almost instantly. The other forward gunner, David (the last names of Task Force 160 were withheld by their request), was hit in the right leg but continued to sweep the terrain with fire.

The pilots struggled to abort the landing but with the loss of an engine, Razor 01 slammed into the ground, its nose pointing up the hill toward the main enemy bunkers. It impacted hard enough to send the Rangers and aircrew sprawling across the floorboards. David was stretched out on the floor, attempting to apply a tourniquet to his bleeding and broken leg with the nylon cord that was normally attached to his 9-mm pistol.

Rounds shattered the cockpit glass, along with the lower leg of one of the pilots, Chuck. Kicking open his emergency side door, the pilot flopped onto the snow alongside the downed bird. The second pilot, Greg, had a chunk of his left wrist ripped off as another round passed through his thigh. Wrist spurting blood even as he held it, the pilot staggered out of the cockpit toward the rear of the aircraft.

Insulation rained as confetti from the rounds peppered all about. Through the right forward window an RPG streaked to impact against an electronic console that started to burn. The aircraft medic, Sergeant First Class Cory, took three rounds to his Kevlar helmet, sufficient enough to knock him down but inflicting only a small laceration to an eyebrow. Though the fire was soon put out, the air was still laced with smoke and bullets, and the enemy seemed to be firing from everywhere. Razor 01 was in a hornet's nest.

In the rear of the downed MH-47E, the rear door gunner and a Ranger opened fire, killing one al Qaeda. But the Rangers knew that they'd all be dead if they didn't get out of the Chinook quickly. With every second counting, a well-rehearsed exit out of the rear of the aircraft to assigned positions around the helicopter was replaced by a desperate evacuation to get out of the hulk that had now become a virtual magnet for bullets.

In the mass exodus, the Chalk 1 Rangers braved the fires and rushed down the rear ramp. Specialist Marc A. Anderson, 30, due to be discharged from the Army within a few months, his term of enlistment completed, was killed while still in the cargo bay. Sergeant Bradley S. Crose, 22, and Private First Class Matthew A. Commons, 21, were both killed on the ramp. To have four men outfitted in full battle armor cut down in such a brief time span only indicated the incredible fusillade of fire they all faced at that moment.

Others were more fortunate. Specialist Anthony Miceli had only his 5.56-mm M-249 Squad Automatic Weapon (SAW)—a light machine gun—struck by enemy fire as he made his way down the ramp while Staff Sergeant Joshua Walker had a bullet stopped by his Kevlar helmet.

From a Predator drone—an unmanned reconnaissance aircraft— above, visual imagery was transmitted back to the commander of U.S. ground forces in Afghanistan, Army Major General Franklin L. "Buster" Hagenbeck, headquartered at Bagram air base just outside Kabul. There, he and his staff watched in stunned amazement the unexpected ferocity of the ambush as the real-time imagery streamed in. Noted Hagenbeck, "It was gut-wrenching. We saw the helicopter getting shot as it was just setting down. We saw the shots being fired. And it was unbelievable the Rangers were even able to get off that [helicopter] and kill the enemy without suffering greater losses."

Laboring through knee-deep snow, the surviving Rangers spread out separately in different directions and returned fire at two enemy concentrations 50 to 75 yards above them on the ridgeline. Each man sought whatever cover he could find among the rocks. From the left rear of the aircraft, at about the 8 o'clock position—the nose of the aircraft pointing at 12 o'clock—the first Ranger out, Staff Sergeant

Raymond DePouli, spotted two or three of the enemy. As tracers impacted on his body armor, he let loose a full magazine of rounds from his M-4 assault rifle that quickly eliminated the source of the firing.

Behind a boulder located under a tree off to the right at 2 o'clock, another cluster of enemy fighters was engaging with machine guns and RPGs, one of which impacted and detonated near the right side of the bird. Shrapnel whizzed through the air, wounding Staff Sergeant Kevin Vance in the left shoulder, Specialist Aaron Totten-Lancaster in his right calf, and Self in the right thigh.

Another RPG followed, deflecting off the MH-47's tail. Moving around from the far side of the aircraft, DePouli could see the man who fired the RPG and shot him in the head.

The sounds and smells were different than many of them had expected. For Self, it all seemed surreal. "You see something happening and it doesn't seem real. We understood we were getting shot [at]. But it just seemed like a bad movie."

The fright and disorientation of the first moments of intense combat were soon replaced by a sense of anger and outright indignation. Walker vented his irritation with a shout, "Who do these guys think they are?" followed by a rush forward. Laying down fire from his M-4 as he advanced, Walker moved to the aircraft's right side and took up a position behind a rock. He was soon joined by Self and Vance.

Totten-Lancaster, who'd been struck in the right calf by shrapnel from an RPG, attempted to join the other three Rangers, only to realize that he couldn't stand on his right leg. The wound did not stop him as he rolled the several yards necessary to reach the rock.

A bit farther behind and to the right of the small group as they scanned the ridge above them, DePouli and Private First Class David Gilliam—the newest member of the platoon and the only one not to take a hit to the body or equipment—sought cover behind another rock, where they found the bullet-ridden body of an al Qaeda and an unused RPG. The seventh surviving Ranger of Chalk 1, Miceli, remained on the left side of the aircraft, guarding that flank with a team member's weapon that he'd grabbed off the ground to replace his damaged SAW.

For the next few moments, a grenade exchange took place with several Rangers attempting to hurl them towards the closest enemy position approximately 50 yards away while a similar effort was attempted by some al Qaeda. The exchange of explosive charges proved to be ineffective for the farthest the grenades could humanly be thrown was only about 35 yards.

Inside the helicopter, the wounded were being attended to in the shot-up cargo bay by Cory and the two para-rescue jumpers (PJs), Tech Sergeant Cary Miller and Senior Airman Jason Cunningham. All three wounded were crew members: Chuck—who'd been pulled around to the back of the aircraft after escaping through his side cockpit door—with a serious leg wound, Greg—who had to have a tourniquet applied to stop the bleeding of his left wrist and David, who still had the tourniquet applied to his leg wound.

Electing to keep the wounded inside the aircraft in order to shield them from enemy fire, Cory and the two PJs were compelled to crawl on their stomachs within the ruins of the chopper to make their way around, for the enemy could see into the heavily damaged right side of the helicopter from their elevated vantage point off the nose of the aircraft and shot anything that moved inside. Cory also knew that he needed to do his best to shelter the seriously wounded men from the cold; anyone who loses significant amounts of blood is more susceptible to hypothermia.

Outside of the chopper, Captain Self needed help to keep fighting and enlisted the aid of two aircraft crew members, Don and Brian. Returning to the chopper, the crewmen came back with additional ammo and Commons' M-203 grenade launcher that he'd been carrying when cut down on the ramp. Multiple returns to the chopper under the covering fire of the Rangers left both men near exhaustion because of the high altitude. But, they had accomplished their mission, and the Rangers were now ready to carry the fight to the enemy, just as they'd been trained to do.

Unfortunately, given the circumstances of a downed helicopter, wounded, dying, and dead buddies, there was only one offensive choice available to them. To their left was high ground, movement along which would leave them exposed for too long to enemy fire. To their right, the terrain dropped off steeply, leaving no possibility for maneuver. Self said: "That's when we made the decision that the only way to assault would be straight at them."

Gilliam, armed with his M-240B heavy machine gun was told to provide cover fires with Brian serving as his assistant gunner (AG). With Gilliam laying down withering 7.62-mm machine gun fires, Self, DePouli, Walker, and Vance charged. Weapons ablaze and grenades in hand, they assaulted up the hill, laboring through three feet of snow.

Suddenly, after approximately twenty-five yards, halfway up the hill, Self spotted an enemy fighter in a chest-deep entrenchment. Glancing around a tree, the enemy took a shot before disappearing down into his hole. The reality of the situation struck Self immediately as he realized that he and his three fellow teammates were attacking bunkers shielded by logs and camouflaged with branches, leaves, and snow. They did not have sufficient manpower or weaponry for such a task.

"Bunker! Bunker! Bunker! Get back!" he shouted to the others.

The small team returned to the rocks. The Rangers knew that any further attacks would require more than four soldiers.

Staff Sergeant Gabe Brown, in the meanwhile, had made his way about 25 yards down the slope to the rear of the helicopter to set up a communications post behind a rock from where he was able to finally establish communications with the SEALs. It was the first news Self had that the SEALs they had come to rescue were no longer even on the ridge top! They had already moved some distance down the mountain even before the Rangers had arrived.

With rescuing the SEALs no longer an urgent matter, Brown went to work attempting to contact any U.S. fighter bombers in the area for close air support (CAS). His frustration began to mount as communications glitches continued to hamper his ability to contact any support. At last, nearly 20 minutes after the helicopter had been shot down, he was able to reach Bagram and receive the additional radio frequencies he needed to talk to the inbound fighter-bombers, two F-15 Eagles and two F-16 Falcons.

Multiple gun runs by the CAS with 20-mm cannon fire failed to destroy the bunker. Finally, despite the danger and closeness of the target to the Rangers, the aircraft dropped three 500-pound GBU-12s (Guided Bomb Unit).

The first bomb impacted down the hill, behind the helicopter, prompting Brown to chide the pilot, just a bit.

"Whoa, you almost got us with that one. Can you move it a little closer to the tree?"

The second strike was on the ridge crest, to the front of the downed bird. The third—and last—strike proved to be a direct hit on the bunker, destroying the tree before it, in the process.

Though the air strike had suppressed the enemy fire and destroyed one of their bunkers by mid-morning, the high ground of Takur Ghar ridge was still in enemy hands, and Self was no closer to securing it. Enemy fighters could be observed in the distance moving to envelope them from the rear. Spotters in the area and aircraft overhead noted other signs of enemy reinforcement efforts.

With the bunker destroyed, Cory was able to shift his patients to an area behind Razor 01. Though their bleeding had stopped and their vitals remained stable, all three were suffering from life-threatening injuries and were in dire need of quality medical care. They needed to be evacuated immediately.

An apprehensive aircrew unaccustomed to ground combat grew more alarmed about the condition of their three wounded comrades and pressed Self to mount a new assault on the fortified defenses to clear a way for an evacuation. Their alarm grew as an unexpected and unpleasant battlefield situation soon developed when al Qaeda mortar rounds began to fall around the remains of Razor 03, landing in front of the aircraft's nose, then in the rear, and finally down the hill, bracketing the group.

Self shared their sense of urgency and realized that they may all be in trouble soon, if he didn't get his team to the top of the ridgeline. He wished that there were an option but . . ." [The aircrew] didn't understand the timetable that we were really on. They expected things to happen quick, quick, quick. 'You guys run up there and kill the enemy.'"

For a moment, the Ranger captain had second thoughts about his decision to break off the first assault on the northern bunker and withdraw but, in the end, he knew it had been the right decision at the time. The four Rangers assaulting the bunkers were the four most senior combatants on the mountaintop. Had they died in the attempt, the survivors of Razor 03 would have stood little chance against an attack. Self's reply to the aircrew's request to attack was "No."

While Self shared their sense of urgency, he also knew another frontal assault was not an option. To take the hill eventually, the Rangers had to stay put and continue waiting for additional help. They needed reinforcements.

Razor 02 had loitered, circling over the valley, while Razor 01 had disappeared up the mountain. Shortly afterwards, communications with Razor 01 had been lost. Not aware that Razor 01 had been shot down, Razor 02 was eventually directed to put down at Gardez and await further instructions.

Time passed and the Chalk 02 crew's anxiety grew as they waited further word about the fate of Chalk 01. The agonizing wait created frayed nerves. Staff Sergeant Canon, the ranking Ranger in the Chalk, grabbed one of the airmen.

"At one point, I had a crew chief by the collar. I'm screaming at him that regardless of what happened, the first bird only had ten guys on it. That's the bare minimum package. If something happened to them, they need us. We complete the package!"

Time passed and the Chalk 02 crew's anxiety grew as they waited further word about the fate of Chalk 01. Finally, with confirmation that Razor 01 had gone down, Razor 02 was back on its way towards Takur Ghar, a half hour to hour following notification, with the ten Rangers of Chalk 2 and an additional SEAL who had joined the group.

With HLZ Ginger no longer an LZ option, Razor 02 searched for an offset alternate LZ.

Ray, the pilot, noted that, "It's the side of a mountain, so there are not a whole lot of places to land. You basically hunt and peck around."

It was 8:30 AM before they were able to find a space clear enough to touch down with all four sets of wheels—two sets, four wheels up front, two sets, two wheels in the rear.

The Rangers were informed by the aircrew that Chalk 1 would be straight ahead, 250 to 300 yards away. Astonishingly, after a quick glance around, Canon realized that they were way off the mark. In reality, they were 800 yards to the east and 2,000 feet below their trapped fellow comrades!

Though some of the Rangers knew a thing or two about the cold and snow and all had some basic experience as a result of their Ranger training when it came to mountain climbing, none of them was prepared for the task they had before them: a climb in thin mountain air, weighed down with weapons, ammo, body armor, and equipment along a towering, forbidding 45 to 70 degree slope that was mostly covered by three feet of snow.

Before they even began to move out, they observed about 1,000 feet up to their right a small cluster of Americans, two of them wounded, slowly making their way down the mountain. They were the SEALs that the Rangers of Chalks 1 and 2 had been sent to rescue.

The SEAL who had joined Chalk 2 requested that the Rangers assist the SEAL team prior to commencing their climb. The request was transmitted by Canon to Self, who immediately declined.

"I've got casualties up here, and I need you now more than they need you."

The SEAL and Rangers finally went their separate ways, each moving to assist their own—he across the mountain to join his fellow SEALs, Chalk 2 up the mountain to join their fellow Rangers.

"It was kind of like a merry-go-round," observed Chalk 2's Ranger medic. "We were trying to go up and they were coming down."

Later, the SEAL team would acknowledge that it would not have lasted much longer if the Rangers had not arrived when and where they did. Though they had already started to egress down the mountain, they were still under fire. The arrival of Chalk 2 diverted those fires.

Being a remote mountain under snow, there was no trail to follow, so the Rangers had to make one of their own. Despite not knowing what lay ahead of them, the Rangers did know one thing: time was undoubtedly running out on the small group of survivors above. They had to move quickly.

"Quickly," however, proved only to be a relative term as the Rangers struggled to find traction on the loose shale rock of the mountain's steep slope.

"Just the grade of the ridge made it an unbearable walk, never mind the altitude. It was enough to where my guys' chests felt heavy and my joints were swollen," Canon was to comment later.

Periodically, the Rangers found it easier to crawl along on their hands and knees. Though sporadic and poorly aimed, enemy mortar rounds periodically fell around them, reminding them not to tarry.

Team leader Sergeant Stebner would spur the men on whenever they stopped to look and see where the rounds were coming from.

"You can't stop. It's not going to do us any good to stop. We have to keep moving."

<hr />

Compounding their difficulty, the Rangers were improperly outfitted with their equipment and clothing. When first alerted for the mission, most of the team members were under the impression that the operation would be a standard quick in-and-out rescue with little maneuver or movement involved. With the expectation of long waits in drafty, cold choppers or in stationary positions forming a defensive perimeter on the ground, many of the Rangers had dressed in thermal underwear under bulky parkas. Consequently, while dressed for the cold, they were beginning to experience heat exhaustion as those wearing the heavy gear began to sweat profusely from the exertions of their climb. Others had made the mistake of foregoing their Gortex or cold-weather boots for their suede desert boots that were now taking a beating and sopping up the wetness of the snow like sponges.

Furthermore, though Rangers were classified as "light infantry," there was nothing light about the soldier's load they had to "hump." In addition to their uniform and outer garments, they wore twenty-two pounds of body armor and a four-pound Kevlar helmet. Each man also carried a weapon—the majority an M-4 assault rifle, seven to twelve magazines of ammunition—each with thirty rounds of 5.56mm, two to four grenades, a 9-mm pistol with at least two clips of fifteen rounds each, knives, flashlights, communications gear, night vision devices, a first aid kit and three quarts of water—weighing an additional six pounds. All told, these "light infantrymen" carried a minimum of eighty pounds on them.

Heavy machine gunner Specialist Randy Pazder carried the heaviest load of all. His M240B weapon alone weighed twenty-eight pounds. Just with his weapon, thirty pounds of ammunition, and his protective armor, Pazder was already well maxed out at eighty-four pounds, seventy pounds being the recommended maximum load. His assistant gunner, Specialist Omar Vela, carrying an additional thirty pounds of ammo, a spare gun barrel, and his own personal weapon and ammo was not too far behind.

Slowly, the Rangers trudged up the mountain, at times throwing their weapons ahead into the snow then climbing forward to pick them up. If they appeared to tire, Canon would urge them forward.

"You need to get to the top of the hill, where we'll be in a static position and can rest. We've got to go, our guys need us."

Approximately 1,000 feet up, with another 1,000 feet to go, the Rangers shimmied around a rock to lift themselves past a tree that protruded from the mountain face. Momentarily stopping, Canon took a good, hard look at his exhausted men.

"Everybody had the, you know, 'Man, this sucks' face—just a long face."

It was time to make some adjustments.

As the Rangers shed their heavy clothing, Canon contacted Self and received authorization to lighten their loads by allowing those who wished to discard the six pound ceramic back plate—with a cost of $527—that supplemented the basic Kevlar vest. Ballistically, the plate was designed to stop anything from a 9-mm round from a pistol or a 7.62-mm round from a high-powered AK-47 assault rife—the type used by al Qaeda. Though the load lifted from their backs was relatively minimal, it did allow for greater comfort and mobility.

To avoid their falling into enemy hands, Canon had his men shatter them by tossing the plates to the distant rocks below.

"It's the most expensive Frisbee you'll ever throw."

Rucked up and "refreshed," their thoughts once again turned to their buddies in need. The climb resumed.

Canon crested the final rise around 10:30 AM, surprised that the climb, estimated to take 45 minutes, had taken two hours. More startling, however, was the carnage. At the rear of the downed aircraft, they saw their comrades spread about. Though Canon and his men expected to see casualties, they did not expect to see so many dead and wounded comrades. But, there was neither time for reflection nor for rest for Chalk 2. Finally, Self had his reinforcements; they would take the mountaintop.

The Rangers task organized into two groups. The support-by-fire element consisted of two heavy machine gun teams, Chalk 1's Gilliam and his assistant gunner, Brian, and Chalk 2's Pazder and his assistant gunner, Vela. Canon positioned himself between the two gun crews to control their fires. In close proximity near a rock beside Razor 01, it would lay down massive firepower at the right time.

The assault team was composed of six Rangers from Chalk 2: Sergeant Eric Stebner, Staff Sergeant Harper Wilmoth, Sergeants Patrick George and Walker, and Specialists Jonas Polson and Oscar Escano. Its attack position would be located behind another rock, to the left and slightly forward of the machine guns that would provide the supporting fires.

The primary objective of the assault was an enemy bunker situated 50 yards to the right of the nose of the helicopter. Though a CCT directed airstrike minutes before had appeared to silence it, there was no certainty that al Qaeda had been killed, reoccupied the position, or had fled. The only way to be certain was to close on the position, destroy the enemy, and seize the terrain. And that was exactly what the Rangers intended to do.

At 10:45 AM, the machine guns opened fire, red tracers streaking forward, lapping at suspected enemy positions and forcing al Qaeda to remain under cover. Moving out in the "high ready"—weapons shoulder high, eyes over the gun sights—the six Rangers began to storm the hill as quickly as they could in the knee-deep snow, feet slipping on the rocks hidden beneath, firing and tossing grenades as they advanced.

Trained to conduct an assault with two four-man teams—one for maneuver, the second providing immediate supporting fires—and given the difficulty of the conditions, all six served as a maneuver element, plodding steadily forward. The Ranger fires were so heavy that it caused alarm with the aircrew, prompting them to yell out to the Rangers: "Slow down, you're going to run out of ammo."

To the left of the bunker, 40 yards away from the downed Chinook helicopter was a boulder. First to reach the boulder was Stebner, who stumbled across a dead American lying face down in the snow. He didn't know it at the time but he had found Petty Officer 1st Class Roberts.

Only the al Qaeda knows exactly what happened to Navy SEAL Neil C. Roberts, 32, during those early morning hours, alone and in the midst of a merciless enemy, fighting for his life as the sound of Razor 03's beating blades receded down the mountain, into the valley below. Forensic evidence seems to clearly indicate that he not only survived the fall from the helicopter, but that he was also able to activate an infrared strobe light signaling device and engage the closing enemy with his SAW for at least the first half hour as he moved around the LZ. Later, Roberts' SAW would be found near his body with blood on it. There were other indications that he had been able to fire it. Ammunition remained in the weapon, perhaps suggesting that it had jammed.

Following an overhead observation of Roberts being surrounded by the enemy, a Predator drone arrived to provide a video image of the area just as Roberts' strobe light went out. Drone imagery, though fuzzy, possibly showed Roberts finally being captured by three al Qaeda and being led away to the south side of Ginger where the group disappeared into a tree line, approximately fifteen to twenty minutes before the arrival of Razor 04 and its team. All told, evidence suggests that he was able to hold the enemy at bay for an hour, maybe even two, before he was either executed or shot at close range.

Stebner was joined at the boulder by Wilmoth, George, and Escano who immediately rushed the bunker approximately five yards away. Inside, amid the debris of the earlier airstrike, the Rangers found the body of the second missing American, Technical Sergeant Chapman, sandwiched between two dead enemy fighters.

Outside the bunker, Stebner and Polson maneuvered left of the bunker then circled right, firing on the last enemy positions they saw just over the crest. Their rush would prove to be the last as they seized the crest of Takur Ghar.

A sweep of the mountaintop showed what American intelligence had failed to detect, a network of enemy positions dug behind rocks and next to trees, connected by shallow trenches, even a canvas tent sheltering one of the positions. Throughout the area, sheaves of RPGs and Chinese-made 30-mm grenade launchers were strewn about. Also amidst the rubble was a Russian-made DShK heavy machine gun and a 75-mm recoilless rifle. Assorted small arms and long belts of machine gun ammunition lay scattered.

Eight enemy bodies were accounted for on the mountaintop, with one found to be wearing Roberts' jacket. It was 11 AM and the Rangers now held the mountaintop.

Finding the mountaintop a more defensible position, Self began to gather his Rangers and the rest of his team members along the ridgeline. Taking a seat beside Self, Canon listened as Self informed him of the Ranger dead. The three names struck the Ranger sergeant pretty hard but Self knew that he could not afford to have his most senior non-commissioned officer—or anyone else for that matter—preoccupied with the dead men, given their current situation.

"Arin, there's nothing we can do about it now. Let's get the rest of these guys out of here alive, and we'll deal with what we have to deal with when we get back."

Plans were made to begin moving the six dead and three seriously wounded men from their current location behind the helicopter to what appeared to be a suitable pick-up zone (PZ) on the far side of the crest. The movement of the wounded up the steep slope, however, was a difficult and slow process for it took four to six men to move each man or body.

Down in the chopper, Greg's condition was growing worse and he started to speak to Cory as though he were preparing to die. Vance, in contact with his tactical air controllers back at headquarters, was pressed to relay whether or not the PZ was "cold"—meaning clear of enemy fires—and how many of the wounded would they lose if their rescue were delayed. Not sure, Vance directed the question to Cory.

"If we hang out here, how many guys are going to die?"

Cory's reply was at least two, if not all three.

Vance passed along the situation report.

"It is a cold PZ, and we are going to lose three if we wait." The time was approximately 11:15 AM.

The enemy's counterattack arrived with the swoosh of an RPG overhead and the crack and whiz of small arms fire filling the air from three or four of the enemy who were located on a knoll approximately 400 yards to the south of the downed bird. Trees were splintered and pine needles fell all about as bullets ricocheted off the rocks around the feet of those carrying David, the first of the casualties to be moved up to the ridgeline. Forced to drop the litter, the men scattered for their lives. While David lay on his back, exposed to enemy fires, Stebner dashed out twice in an attempt to drag him to cover behind some rocks. Each time, though, he was driven back by heavy fires. Finally, on the third attempt, fifteen minutes or so after the engagement had started, Stebner was able to reach the airman and drag him to safety.

Below, behind the Chinook, Cory and Cunningham had just finished inserting a fresh IV into Greg when a stream of bullets ripped into the exposed casualty collection point. Caught in the open with two seriously wounded men, Cory realized that "we were just going to have to sit there and shoot it out with them. Neither Jason nor I were going to leave."

The fires intensified, an RPG coming straight at them rocketed over their heads, its fuse detonating the explosive warhead over the helicopter. A bullet struck just a few feet in front of Cory, spraying snow upon him. As they continued to engage the enemy, Cory thought to himself, "I have only two magazines left—something has to happen here pretty soon."

In an effort to make something happen, Cory began to crawl up the hill on his stomach for approximately five feet to within six feet of Cunningham before stopping for a moment to turn on his back and return fire. Moments later, both men were simultaneously cut down. Cory was struck in the abdomen by two bullets that hit his ammunition pouch and belt buckle. Unsure as to the extent of his injuries, the medic hesitatingly reached down then pulled his wet hand back. Water, from a punctured canteen. His relief was short-lived, however, for the medic had been seriously hit. Cunningham, also, had been hit, leaving him bleeding profusely from a serious wound to the pelvis area. Though conscious and coherent, he was in considerable pain.

Farther down the hill from the downed chopper, a small group of Rangers, which included Canon and DePoli, was positioned on a perimeter to provide security for the casualty collection point. Opening fire on the enemy position with a heavy and light machine gun, several assault rifles, and a grenade launcher, they could observe some of the enemy maneuvering around the hilltop.

To the left, Pazder caught sight of an al Qaeda popping up. The Ranger quickly killed him with a burst of heavy machine gun fire.

To the east, approximately 800 yards away down a slope, the Rangers spotted a small group of four or five enemy fighters making their way up. While they were far away, they were still within the maximum effective range of the M240B machine gun. However, they were low on ammunition so Canon dispatched Vela back to the chopper to grab some more.

Vela dashed off towards the helicopter located 150 to 200 yards away. Arriving safely, he gathered some belts of 7.62-mm ammunition and began his return run. An eruption of enemy fire forced him to dive behind a rock for cover where he found Stebner, who quickly informed him, "You might not want to be by me because for some reason the enemy doesn't like me."

"What are you talking about?" asked Vela, unaware of Stebner's two failed attempts to extract the wounded Dave, who was still lying out in the open on the stretcher.

In punctuation of Stebner's comment, an RPG whizzed over their heads.

"That's one thing I'm talking about. Every time I get up and move, they shoot at me. And now I'm laying here and they're shooting at us."

Heeding the advice, but still with a task to accomplish, Vela scampered to another rock outcropping where he joined DePouli. Wrapping the machine gun ammunition in the spare gun barrel bag, Vela attempted to toss it across the remaining exposed distance to Canon on the far side. The throw was not strong enough and the bag thudded to the ground, only halfway across. Scurrying out on all fours, Canon dragged the bag back.

Given that Canon had the better angle to shoot, Pazder passed him his machine gun. Opening fire, Canon placed plunging fires down into every tree or bush where an enemy might be hiding. The Rangers never saw an enemy at that location again.

As the fight continued to develop below, Self on the ridge realized one thing. Had his Rangers of Chalk 2 not made the valiant effort and climbed the mountain when they did, the small band of Chalk 1 survivors would now have been caught between the enemy's plunging fires from the ridgeline and their fires from the distant knoll into their exposed rear. Literally and figuratively, they would have been between "a rock and a hard place" and most likely would not have

survived. In the end, Chalk 2's arrival had proven to be most fortuitous—and timely.

_____

CAS finally arrived back on station in the form of Navy F-14 Tom-
cats around 11:45 AM. On the mountaintop, Brown began to call the
strikes in on the knoll below and a succession of 500-pound bombs
was dropped. Shrapnel streaked through the air with each detonation. Noted Brown, "We could see the bombs go down the hill below
us, and we heard the material rising up past us, whizzing through
the air." After one blast that forced DePouli's Kevlar helmet back on
his head, the sergeant called Self on the radio.

"Can we get a little bit of a 'heads-up' [notification] down here
the next time we're going to make a bomb run like that?"

"Yeah, sure, no problem," was the platoon leader's understated
reply.

Six bombs later, the enemy on the southern knoll had been eliminated, and the transfer of casualties to the other side of the ridge crest
could begin.

_____

Self was now faced with an even greater sense of urgency regarding the evacuation of his casualties for he now had two freshly
wounded men, Cory and Cunningham, in addition to the three others. Worse was his realization that, while Cory's wounds did not appear life threatening at the moment, the same could not be stated for
Cunningham's injuries.

_____

Though the Ranger medic was able to staunch the para-rescue
jumper's external flow of blood, he had no idea just how bad the
internal wounds were. Coincidently, Cunningham had strongly lob-
bied and convinced his command just days prior to authorize the
inclusion of blood packs in their PJ kits. He was now the first benefi-
ciary of his own common sense.

Despite their concern over the number of casualties designated "urgent surgical," the senior command was even more worried about another daylight rescue attempt. Further complicating matters was Operation Anaconda which was still ongoing, with nearly 1,400 American soldiers maneuvering on the ground and a multitude of U.S. aircraft streaking across the skies throughout the valley.

To complicate matters further, intelligence reports had indicated as many as seventy al Qaeda converging on the Takur Ghar mountaintop. However, given the intelligence community's earlier failures at the start of Anaconda, these reports should have been considered suspect from the beginning. This was confirmed later by the observations of a combat air controller, Air Force Technical Sergeant Jim Hotaling, located at an observation post two miles to the south who had "eyes" on Takur Ghar. While he did observe several very small groups of al Qaeda maneuvering in the area throughout the day, he saw nothing close to the seventy reported.

"Most of the enemy I was engaging," Hotaling would later report, "was a good 1,500 to 2,000 yards away from their position, down on the bottom of the mountain and in the creek beds."

Even if there were truly seventy al Qaeda, were we missing the fact that we, the United States, were *the* only super power with mastery of the air?

Vance consistently badgered his headquarters, informing them that it was safe to send in a MEDEVAC.

"I kept telling [the] controller that we lost another [wounded soldier], [that it was a secure] PZ, [and asking] when are we getting exfiltrated? Controller said to 'hold on.'" After asking him three times, PL [platoon leader—Captain Self] expressed urgency at getting the team out of there. I continued to tell controller but he just kept telling me to hold on. After the third time, I handed the [radio] hand mike to the PL and asked him to tell controller the same thing. I tried to keep a monotone voice. There were times that I tried to throw some words in there to make controller realize that we [had] to get out. It became a personal conversation, and we kept saying we have to get out of here."

Finally, the Ranger medic got on the radio to emphasize the urgency of the men's wounds.

"I felt as though if I started making a big deal about their conditions, then it would worry my patients. You want to be open and honest, and I was, but I wasn't jumping up and down, ranting and raving, that this guy was going to die if we don't get him off this mountain. I said, 'Listen, here's the story. I've got two urgent surgical patients, and we need to be evac-ed.' And their response was, 'Roger, we understand.'"

Even Self's urgent appeals for an evacuation were rejected by his commanders, who refused to authorize any additional attempts during daylight.

While the Ranger medic did his best to assure the wounded men repeatedly that help was on the way, the aircrew—the pilots in particular—knew better. It was daylight and they had already lost one bird.

The medic would later recall, "I kept coming back to them saying, 'Hey guys, listen, they're going to come get us, we're going to be out of here soon, hang in there.' And it was the helicopter pilots who were pretty upfront about it, and they said, 'We know we're not leaving until dark because that's just the way it is.' I knew in the back of my head that the chances of them coming during daylight hours were slim to none, but I was trying to be positive about it."

And what was seriously wounded Jason Cunningham's reaction to all this? "For the most part, he listened," the medic reported.

By around 5 PM, the sun began to sink beyond the ridgelines and it began to grow dark. Along with the fading light, the temperatures began to drop, and the wind began to pick up as the ridgetop turned bitterly cold.

From the mountaintop, Stebner and Wilmoth admired the stunning vistas as the sun set. How strange it was, Stebner would mention to Wilmoth, to be in such a beautiful place amid such brutal conditions.

Momentarily, Self, also, became somewhat reflective, thinking back to a passage from Psalm 121 that some of the Rangers had read during a Bible study group at Bagram the evening prior to the mission. It had been a favorite of his since his days as a cadet at West Point.

"I lift up my eyes to the hills, where does my help come from?"

But contemplation and mourning would prove to be too self-indulgent for now.

"There were a few times here and there where guys would start to reflect on what had just happened, and their minds started to affect them a little bit," Self recollected. During those times, he would tell them, "Hey, you've got tomorrow and the rest of your lives for that."

Conditions continued to worsen on the mountain. Breathing became more labored in the dry, thin air. Throats became sore and bled from their dryness leading to nearly everyone coughing up some blood. Dehydration began to set in for many. The chopper was searched for anything and everything that could provide warmth, food, and water. Enough crackers, Power Bars and Meals-Ready-to-Eat (MRE) were found to assuage everybody's appetite just a bit, but they failed to exacerbate anyone's thirst.

Hypothermia was setting in and an effort was made to keep the wounded men warm. The aircraft's sound insulation liner was stripped and the wounded placed on it to insulate them from the cold ground. Anything that could retain heat—sleeping bags, blankets, jackets, sweat shirts, pants—was placed on top of them, leaving some literally buried beneath a foot of clothing. Others constructed a lean-to out of branches and wood from a shattered tree to act as a windbreaker.

Despite everyone's best efforts, Senior Airman Jason Cunningham, 26, died of his wounds shortly after at 6:10 PM. He had been a para-rescue jumper for only eight months. Such was combat on the slopes of a mountain at 10,000 feet.

The first of the extraction birds from the 160th SOAR arrived at 8:15 PM, approximately two hours after Cunningham died. The black shape of the first MH-47E touched down on the designated LZ, inadvertently positioning its tail ramp in the opposite direction of the Rangers waiting on the perimeter of the LZ with their four seriously wounded comrades. Once again, the Rangers were forced to carry these men an additional seventy-five feet or more through the snow and over ice encrusted rocks to the rear of the chopper.

"[More] than once we had to stop and set down, or one guy slipped on the ice," the Ranger medic would later report. "We never dropped a casualty. But I know it was uncomfortable for the wounded, even with the pain control stuff they were given. I know they were hurting. They made it pretty vocal."

While a MH-47E flew a couple of thousand feet below the peak to pick up the SEAL team, the second aircraft arrived at the ridgetop to evacuate the seven killed in action (KIA).

Petty Officer 1st Class Neil C. Roberts — Navy SEAL

Sergeant Bradley S. Crose — Army Ranger, Bravo Team Leader

Sergeant Philip J. Svitak — Army 160th SOAR, flight engineer

Technical Sergeant John Chapman — Air Force Combat Control Team (CCT)

Specialist Marc A. Anderson — Army Ranger, M-240B machine gunner

Private First Class Matthew A. Commons — Army Ranger, M-203 grenadier

Senior Airman Jason Cunningham — Air Force Para-rescue Jumper (PJ)

A third chopper loaded the remainder of the team and aircrew, seven of whom were the walking wounded. Within an hour, all Americans—the living and the dead—had been loaded and were on their way back to Bagram from what would later be honorably re-ferred to by many in the SOF community as "Roberts' Ridge." The significant error in intelligence analysis that led the American com-mand to believe that Takur Ghar was uninhabited had resulted in the highest altitude battle in U.S. military history.

OBSERVATION

Needless tragedy and loss of life occurred on Roberts' Ridge. Inex-plicably, the lessons that should have been learned from Task Force Ranger's experience in October 1993 failed to be applied and ex-ecuted by the Special Operations Command nearly a decade later in March 2002. The failed lessons ran the gambit from intelligence shortcomings and personnel equipment discrepancies to communi-cations deficiencies, misutilization of airpower and Quick Reaction Force (QRF) confusion.

Here are some comparisons as I see them:

Intelligence

Somalia: Failed to note the sharp increase in the purchase of thou-sands of rocket propelled grenades (RPG) by Somalian warlords on the black market and the weapon's intended use to bring down the Special Operations Black Hawk helicopters over Mogadishu as dem-onstrated by the downing of a 10th Mountain Division Black Hawk prior to the Task Force Ranger battle.

Afghanistan: Failed to note the fortification and occupation of the Takur Ghar Mountain peak, just fifty yards or so above the recon team's insertion point at Helicopter Landing Zone (HLZ) Ginger

even though Takur Ghar Mountain was a strategic position border-
ing the area of operation (AO) of Operation Anaconda.

Equipment

Somalia: Rangers deployed late in the afternoon after hours of delays without the night vision devices (NVD) and water they would
need later as they fought throughout the evening and early morning
hours.

Afghanistan: The Rangers of Razor 02 first found themselves over-
burdened and hampered by bulky cold-weather gear and unsuitable
desert boots they were wearing as they struggled up the mountain to
the crash site only to join with the survivors of Razor 01 and freeze
throughout the remainder of the day as temperatures plummeted below freezing. This mismatch of equipment was further compounded
by the soldier's load that was too heavy a burden to carry at such
oxygen depleted altitudes.

Communications

Somalia: Though communications were satisfactory between ground
and airborne elements, those on the ground—Ranger and Delta—had
great difficulty communicating between themselves for each operated with different systems. This difficulty was further compounded by
the fact that while significant parts of the battle devolved into isolated and relatively individual combat, only the Delta operatives were
outfitted with a personal communications device whereas Ranger
communications only propagated along the chain of command and
not down to the individual member of the squad.

Afghanistan: From the start, the QRF encountered serious communications problems, relying primarily on line-of-sight type of communications rather than SATCOM—satellite communications. Poor
communications led to a critical misunderstanding regarding an
"offset"—an alternate LZ—that resulted in Razor 01 flying directly
into the ambush at HLZ Ginger. Furthermore, there was a complete

loss of communications between Razor 01 and Razor 02 prior to the ambush, even though they had five different communications systems aboard each helicopter. Adding to the confusion, when both teams were finally on the ground they had no ability to communicate with the SEAL Team of Razor 04, the very element they had been dispatched to rescue.

Air Power

Somalia: AC-130H Spectre gunships were located nearby in Djibouti and still under the operational control of the Task Force Ranger commander, Major General William F. Garrison. Though the total flight distance was only 680 miles or so as the "crow flies" and, thus, only a few hours away, their heavy and desperately needed suppressive firepower was not ordered on station—a flight pattern—over the embattled task force.

Afghanistan: Though Operation Anaconda was that conflict's most important engagement, and despite having ample air power available in support of it, too little of it was directed to assist the embattled QRF on the summit of Takur Ghar Mountain, just minutes away by air. When Close Air Support (CAS) was finally provided, the QRF had no laser designation capability to guide precision munitions and had to work through a complicated aiming process that still left them exceptionally vulnerable to their own "dumb bombs" exploding around them.

QRF Procedures

Somalia: Lieutenant Colonel Bill David, battalion commander of 2/14 Infantry and the ground commander of the QRF, was left scurrying not only to learn of the mission Task Force Ranger had embarked upon hours earlier, but, even worse, he and his men were left begging, cajoling and threatening Pakistani and Malaysian commanders to gather a large enough armored force to smash their way into the center of a hostile Mogadishu after learning of Task Force Ranger's desperate straits.

Afghanistan: Though Captain Nathan Self and his men were the designated QRF, Special Operations Command seemed not to have placed as significant a priority on its overall command and control responsibilities, inclusive of critical staff support, as it should have provided. Ordered into the air from Bagram, Razor 01 and Razor 02 flew the 100 miles to Gardez with little situational awareness (SA), much less any situational understanding (SU). That should not have occurred given the aerial observation platforms that were available within the region.

Solutions to these issues are both systemic and command related. Intelligence failures continue to be plagued less by inadequate intelligence gathering and more by poor intelligence analysis—whether it be Somalia, the events of September 11, 2001, Afghanistan, or Weapons of Mass Destruction (WMD) in Iraq. Each was a failure of analysis combined with a lack of vision to perceive our enemy's potentials and capabilities. A greater focus within the intelligence community as a whole may be achieved by an accountability of analysis that are glaringly in error—an action that seems to never take place at any level, tactical to strategic. Too often when a mistake is identified, it is shuffled off to the side or buried under paragraphs of obfuscating explanation. In the future, the intelligence community and our national leaders need to start developing a backbone of character and integrity that allows them to not only identify minor and major mistakes but to also investigate why the mistake occurred and who was responsible for the breakdown.

There are those who would claim that such "finger pointing" serves little purpose other than intimidating and placing fear in those who have failed to perform properly and may, ultimately, stifle the process rather than reform and energize it. In perspective, though, which is the greater sacrifice: to feel intimated or fearful enough to focus on reducing failure or to be dead or mutilated as a result of that intelligence failure? Everyone in the military or intelligence community faces various forms of risk and failure. Some, however, have much more to lose than others. If members of the intelligence community are not up to the task of minimizing failure with insightful and accurate data analysis, then they need to find a new profession.

Unfortunately for now, given the poor history of intelligence fidelity, the consumers of such intelligence need to always view intelligence with a certain amount of circumspection. And, furthermore, soldiers should never place their security in the hands of analysts who will never have to suffer from their own mistakes. Always anticipate and prepare to respond to the unexpected, no matter what an intelligence report states.

Special contingency equipment like cold weather gear and supplies like MREs, power bars, and water, should be kept in small preconfigured loads contained within a few duffel bags that can be easily carried by helicopter. At the objective, these duffel bags can be kicked off the aircraft to be used later in the mission, left behind for future recovery or just discarded. Surely a super power can absorb such a cost?

As for the soldier's load, leaders—good leaders—directly supervise what their subordinates carry. Excessive weight wears down and significantly hampers an infantryman's combat effectiveness. A good place to start when learning how to tailor the soldier's load more appropriately is Lieutenant Colonel Raymond Millen's excellent tactical book, *Command Legacy*.

Given the technological superiority of this nation, the communications shortcomings continually demonstrated at the tactical level, especially by Special Operations Forces, are simply irresponsible—or worse, criminal. If communications is still such an issue with America's most elite forces, imagine the plight of the average ground troop! This nation can communicate with satellites that are hundreds of millions of miles away on the fringe of our Solar System or with Rovers on Mars, and yet two of our helicopters, each carrying five separate communications systems on them, cannot communicate with each other when only miles away in a mountainous environment? There is something seriously wrong with that. It shouldn't take a rocket scientist to fix that with the installation of SATCOM systems within the helicopters, especially those Special Operations aircraft that have the potential to serve as QRF platforms.

Air Supremacy has been a hallmark of American military might since World War II, and this supremacy will most certainly continue in the foreseeable future. With such supremacy, especially in low intensity conflicts such as Afghanistan, there are few if any reasons why continuous close air support (CAS) with precision guided munitions cannot and should not be on call 24/7 for Special Operation Forces that literally operate on the fringe . . . or behind enemy lines. The combination of this CAS and the equipping of the QRF with laser designators for guidance of precision munitions would eliminate the need for ground observers to dangerously "adjust fire" of attacking aircraft. Given the nearby air resources committed to Operation Anaconda and the lack of any similar ongoing life and death struggle as that faced by the beleaguered Americans on Tkar Ghar Mountain, one must surmise that the Special Operations Command (SOCOM) lacked a clear focus of what its true priority should have been at that specific point in time.

In past conflicts, American forces have always been assured that every effort would be taken on their behalf to secure their safety in the face of potential annihilation. Such efforts have even been captured in the film *We Were Soldiers* where Mel Gibson, portraying Lieutenant Colonel Hal Moore—the commander of an infantry battalion about to be overrun by overwhelming North Vietnamese forces—reluctantly calls a "Broken Arrow," a code phrase that indicates that an American unit is about to be overrun and destroyed. [Though this term was misused by Hollywood in the movie—for a "Broken Arrow" is a military code phrase alerting military governors of a nuclear weapons accident/incident—it is the concept that's important.] In response, such a code phrase designation would require the chain of command to commit all available airpower in support of that unit's desperate fight. That would have made a difference on 4 March 2002 in the mountains of Afghanistan. That is what should have happened, and the fact that it did not is inexcusable.

Ultimately, though, the debacle of Takur Ghar was the result of a QRF process that still seems to be as lacking in Afghanistan as it was in Somalia. As critical as it is for Special Operations troops deployed in harms way to feel that "the US Cavalry" in the form of a QRF will ride to their rescue when needed, SOCOM seems still to treat the QRF as an afterthought. As long as this continues, American soldiers such as Ranger Specialist Jamie Smith, in Mogadishu, and Senior Airman Para Rescue Jumper Jason Cunningham, on Takur Ghar, both KIA as a result of wounds that eventually became mortal over time, will continue to be needlessly sacrificed in service to their nation. Certainly, warriors of the world's only Superpower will die in conflict, but they shouldn't die that way. An unnecessary death should be a rare event in our modern army and not a result of inaction on the part of a chain of command sitting safely in the rear watching the action from a high definition flat screen.

How do you fix this problem? The solution is one of priority and command emphasis. There needs to be an approach that spells out accountability and staffing in support of the QRF. This approach and model of organization is found in the Army Field Manual in the form of a "Rear Area Operations Center" (RAOC)—an *ad hoc* organization that is defined but not energized until it is needed. Then it becomes a primary and viable command and staff entity. Nearly every Army field grade officer—the rank of major and above—is familiar with the concept and, thus, should have little problem with planning or execution . . . if it's a command priority.

While Takur Ghar was a failure in command, the comradery exhibited on Roberts' Ridge that cold and brutal March day in 2002 was as exemplary and amazing a display of brotherhood as can be cited in any annals of small unit action. It was a display of brotherhood that can no better be epitomized than by Staff Sergeant Canon's anxious and impassioned plea to launch his Rangers in support of those Rangers aboard Razor 01 after the lead aircraft disappeared.

"The first bird only had ten guys on it. That's the bare minimum package. If something happened to them, they need us. We complete the package!"

We complete the package. No matter what the circumstances, no matter what the adversity, no matter what the danger, Canon and his fellow Rangers knew they had to join their comrades to help ensure their safety, and they would do whatever it took to accomplish that objective. It was a bond of brotherhood linked through shared sacrifice and wrapped firmly by the ties of a creed: "Never shall I fail my comrades; I will never leave a fallen comrade to fall into the hands of the enemy; Readily will I display the intestinal fortitude required to fight on to the Ranger objective and complete the mission, though I be the lone survivor."

In 1959, science-fiction writer Robert A. Heinlein wrote *Starship Troopers,* a work that was more renowned for its social commentary and military leadership philosophy than its thought-provoking futuristic scenario of interstellar war against a "bug" race that lived underground. In the culminating passage of the book, the Mobile Infantry are laying siege to a planet when a platoon sergeant and half his platoon disappear into a large hole that suddenly appears below them. The platoon leader, Lieutenant Johnnie Rico, peers into the vast darkness of the unknown into which some of his men have disappeared and then turns to one of his NCOs, Sergeant Cunha, directing him to take charge of the remaining platoon members—implying that he is to remain above ground and secure the area. As Rico, by himself, jumps into the hole and begins to work his way down a tunnel in search of his missing men, he hears Cunha behind him, issuing orders.

"Section! First squad! Second squad! Third squad! By squads! *Follow me!"* and down into the hole Cunha jumps after his lieutenant.

As many times as I have read this passage, that imagery still elicits a chill down my back and a grunt of approval. Heinlein had it right. He had captured the essence of the Brotherhood within the Profession of Arms in 1959 with the fictional Sergeant Cunha's "Follow me!" In 2002, the reality version of the Ranger Brotherhood was strengthened by Staff Sergeant Canon's "We complete the package!"

"Throughout the war, you were always in my mind. I always knew if I were in trouble and you were still alive, you would come to my assistance."

—Major General William T. Sherman (1820-1891) in a letter to
Lieutenant General Ulysses S. Grant

Ranger preparation for an air assault insertion by CH-47 Chinook helicopters in Iraq, similar to missions in Afghanistan. (date unknown)

MH-47E Chinook Razor 03 heavily damaged by al Qaeda fire during its attempt to insert the SEAL team on Takur Ghar. (Mar. 2002)

Predator drone infrared footage of Razor 1's initial battle
(4 Mar. 2002)

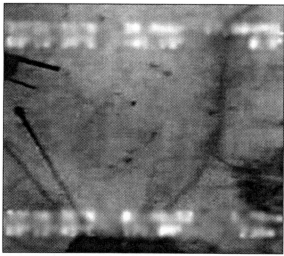

*A rocket propelled grenade (RPG) streaks towards the
downed CH-47 from the lower left.*

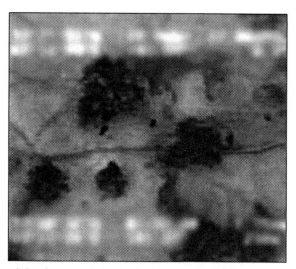

al Qaeda massed on top of Takur Ghar.

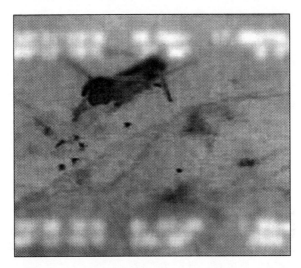

The surviving Rangers of Chalk 1 deploy from the heavily damaged Razor 1 to establish a hasty defensive perimeter to the rear of the aircraft.

Razor 1 downed near the top of Takur Ghar. (Mar/Apr2002)

Still heavily engaged in battle, Rangers and wounded crewmen can be spotted on Razor 1's ramp and to its rear. (4 Mar. 2002)

Specialist Anthony Miceli securing a position on the perimeter. (4 Mar. 2002)

Specialist Randy Pazder, manning his M240B machine gun, overwatchs the Takur Ghar crest. (4 Mar. 2002)

al Qaeda bunker located below the Takur Ghar crest. (Mar. /Apr. 2002)

al Qaeda DShK 12.7mm heavy machine gun that first appeared abandoned but soon proved to be deadly. (Mar. 2002)

NINETEEN

RANGER LESSONS LEARNED

Those who cannot remember the past are condemned to repeat it.
—George Santayana
Philosopher, poet, literary and cultural critic, a principal figure
in Classical American Philosophy, 1863–1952

FINAL OBSERVATIONS

THE TERM "ELITE" THROUGHOUT history has made many a leader cringe for there has always been a price associated with that term, and the cost has invariably been siphoning of the best men and resources from the traditional, conventional, line units. Since the advent of the modern Ranger during the Second World War, the United States Army also cringed as it transitioned through cycles of activating and disbanding such units.

With the formation of the 75th Ranger Regiment in 1984, it seems that the military, as a war-fighting institution, has finally come to grips with the realization that such organizations are not only an integral part of our success on the tactical battlefield but also the cutting edge of our strategic force projection capability. In addition, despite the dedication of men and resources to the 75th Ranger Regiment and the Ranger Training Brigade, there has been a significant "return on investment" regarding the spread of the Ranger ethos and warrior spirit throughout the Army as a whole. These attributes are in keeping with the Chief of Staff General John Wickham's 10 May 1984 directive to Colonel Wayne Downing—the first "modern" colonel of the regiment— "to return these men to the line units of the Army with the Ranger philosophy and standards." That has happened, and the United States Army is a much greater institution as a result of the Ranger philosophy and standards.

Some significant lessons learned from Ranger operations throughout history:

The need for Ranger units

There are those who will argue that there is nothing Rangers do that cannot be done by good conventional infantry. Given the correct set of circumstances, such as full manning, continuity [experienced cadre and unit members who have served with the unit for some time], adequate training, and proper resources [supplies, equipment, and facilities], there appears to be some validity to that position. But conventional infantry seldom operates under the best set of circumstances. Attrition and the routine tasks associated with traditional conventional warfare preclude these infantry units from devoting necessary time and resources to the more specialized operations of the type conducted by the Rangers. Even Darby's Ranger Battalions were unable to fulfill conventional infantry roles and execute specialized missions without sustaining horrendous casualties even though they were better trained and prepared for those specialized missions than any conventional infantry unit would be. If special units such as the Rangers are not held in reserve for specific and unique missions, then when such missions are necessary, units capable of accomplishing them would not be available in time. Given the short notice to execute a task, as experienced during World War II and more recently in Grenada and Panama, only exceptionally well-trained and cohesive Ranger units can successfully accomplish high-risk missions.

The more Rangers look like conventional infantry, the more likely they are to be committed in conventional infantry roles.

As the firepower of lightly armed Ranger units increased with the addition of antitank units and heavy mortars during the Second

World War, the more these units grew to resemble conventional in-fantry units. This "look" led to deployment of Darby's Ranger Force, 1st, 3rd, and 4th Ranger Battalions in areas such as Salerno and the Winter Line in Italy. In all respects, this diluted the Ranger effective-ness, the casualty list grew, replacements grew scarce, they were less trained, and this eventually led to even greater losses over time. Rangers must be employed for missions with operational vice tacti-cal significance. Always!

Ranger units need time to replace and retrain for losses endured.

Inexperienced and untrained replacements dilute and weaken the cohesion and effectiveness of specialty units. During World War II, both the 5th Ranger Infantry Battalion at Zerf and the combined 1st and 3rd Ranger Battalions at Cisterna suffered excessively heavy losses as a result of having to commit untrained and unseasoned Ranger replacements in specialized operations before they were ready. With the establishment of the U.S. Army Ranger School and the dispersion of school- and regiment-trained personnel through-out the Army, there currently exists a dependable source of Ranger-trained replacements should the need ever arise. The funding of this Ranger School is a highest military priority for the health of our total Army effort.

Surprise is critical to the success of light forces.

Ranger units traditionally go into combat only with what they can carry, against well-defended objectives. To attack a well entrenched, alert enemy has to be considered an exercise in suicide such as, for example, the 2nd Ranger Battalion's daring seaborne assault against Pointe-du-Hoc at Normandy. On the other hand, complete surprise such as was achieved by the 3rd Ranger Battalion at Porto Emped-ocle, Sicily, or the 6th Ranger Infantry Battalion Cabanatuan prison camp raid in the Philippines usually results in victory without loss of life.

Full mission rehearsals must be conducted with all involved elements.

The failure to conduct "full dress rehearsals" leads to disasters similar to that experienced by Detachment-Delta and C Company,

3rd Ranger Battalion at Desert One, Iran (U.S. Embassy hostages). The relatively smooth and coordinated operations experienced by the 75th Ranger Regiment during their Operation Just Cause airborne assaults serve as a model for all future joint operations.

Communications that work.

American technology is arguably the finest in the world and almost certainly the finest in history. Yet, individual Rangers are not equipped to communicate with each other as in Somalia. Nor did their Special Operations support aircraft guarantee—with a reasonable degree of reliability—the ability to communicate with each other as in Afghanistan. The technology and capability to do so are unarguably available. Today. The underlying question becomes: how much is a Ranger's life or mission's success worth to this nation?

Intelligence that is accurate.

Intelligence failures are a common theme throughout Ranger history. Ranger units have been wiped out in the fields of Cisterna, Italy. Rangers have been ambushed and nearly annihilated in the mountains of Afghanistan. Rangers have been surprised by numbers, weapons, and tactics of the enemy in the streets of Mogadishu. Rangers have parachuted into combat on Grenada with no idea of the enemy situation. This is just inexcusable. We have to start getting it right or more good men will continue to needlessly die. The defense community must recognize the limitations of technical intelligence. Rather than relying solely on technical means, the Rangers should establish long-range reconnaissance teams to gather and disseminate real time intelligence before the main body deploys. Such organizations are not unprecedented, and the lethality of Ranger missions demands as much ground truth as possible.

Contingency planning and the Quick Reaction Force (QRF).

The Rangers' greatest strengths also create their greatest weaknesses. While stealth and shock may successfully get Rangers to the objective, once the surprise wears off the enemy and he starts to fight back, it's up to the lightly armed infantry to extract themselves from harm's way. When deep in enemy territory, surrounded at every turn by hundreds of potential enemy safely entrenched behind cover and concealment, it is critical to have plans that provide them the combat power they would then lack—protection and survivability. No matter how well trained or how tenacious, light arms, limited mobility and lack of armor protection lead to significant difficulties and casualties, as experienced by Task Force Ranger in Mogadishu or in the mountains of Afghanistan.

Interestingly enough, if one carefully reviews Ranger history, there is not a single example at the team, squad, platoon, company, battalion, or regiment level of any Ranger operation that was not carried out in an aggressive and dedicated manner. Even when directed by higher headquarters to execute missions that were poorly planned or poorly coordinated, the Rangers never shirked their responsibility, even when the odds or circumstances were totally confused or overwhelmingly against them.

By anyone's standards, the Ranger battalions of the 75th Ranger Regiment are the most elite and proficient light infantry in the world. As long as they continue to remain focused on their mission and to keep in mind the lessons learned of the past, the Regiment will remain capable of fulfilling the Army Chief of Staff's initial charter to the Regiment, that of General Abrams, *"of fighting anytime, anywhere, against any enemy, and winning."*

Appendix A

MILITARY AWARD DESCRIPTIONS

COMBAT AWARDS

Note: full award descriptions are defined under Title 10 of the United States Code.

MEDALS AND CITATIONS ARE awarded for exceptional actions in the military, often in combat. Neither in life, nor in battle, does the opportunity present itself so that every man or woman, soldier, sailor or airman can distinguish himself/herself beyond comparison. Though there are varying degrees of heroism as defined by the awards noted below, in a real sense, everyone deserving of a medal can be cast in the same lot; they have gone beyond the call of duty for a purpose greater than themselves.

The *Medal of Honor* (MOH) is the highest medal awarded by the United States. It has only been awarded 3,428 times in the nation's history. The Medal of Honor is awarded by the president in the name of Congress to a person who, while a member of the Army, *distinguishes himself or herself conspicuously by gallantry and intrepidity at the risk of his life or her life above and beyond the call of duty while engaged in an action against an enemy of the United States;* while engaged in military operations involving conflict with an opposing foreign force; or while serving with friendly foreign forces engaged in an armed conflict against an opposing armed force in which the United States is not a belligerent party. *The deed performed must have been one of personal bravery or self-sacrifice so conspicuous as to clearly distinguish the individual above his comrades and must have involved risk of life.* Incontestable proof of the performance of the service will be exacted and each recommendation for the award of this decoration will be considered on the standard of extraordinary merit.

The *Distinguished Service Cross (DSC)* is awarded to a person who while serving in any capacity with the Army, *distinguished himself or herself by extraordinary heroism not justifying the award of a Medal of Honor;* while engaged in an action against an enemy of the United States; while engaged in military operations involving conflict with an opposing or foreign force; or while serving with friendly foreign forces engaged in an armed conflict against an opposing Armed Force in which the United States is not a belligerent party. *The act or acts of heroism must have been so notable and have involved risk of life so extraordinary as to set the individual apart from his or her comrades.*

The *Silver Star* (SS) is awarded to a person who, while serving in any capacity with the U.S. Army, is *cited for gallantry in action against an enemy of the United States while engaged in military operations involving conflict with an opposing foreign force,* or while serving with friendly foreign forces engaged in armed conflict against an opposing armed force in which the United States is not a belligerent party. *The required gallantry, while of a lesser degree than that required for the Distinguished Service Cross, must nevertheless have been performed with marked distinction.*

The *Distinguished Flying Cross* (DFC) is awarded to any person who, while serving in any capacity with the Army of the United States, *distinguished himself or herself by heroism or extraordinary achievement while participating in aerial flight.* The performance of the act of heroism must be evidenced by voluntary action above and beyond the call of duty. *The extraordinary achievement must have resulted in an accomplishment so exceptional and outstanding as to clearly set the individual apart from his or her comrades or from other persons in similar circumstances.* Awards will be made only to recognize single acts of heroism or extraordinary achievement and will not be made in recognition of sustained operational activities against an armed enemy.

The *Soldier's Medal* (SM) awarded to any person of the Armed Forces of the United States or of a friendly foreign nation who, while serving in any capacity with the Army of the United States, *distin-*

guished himself or herself by heroism not involving actual conflict with an enemy. The same degree of heroism is required as for the award of the Distinguished Flying Cross. The performance must have involved personal hazard or danger and the voluntary risk of life under conditions not involving conflict with an armed enemy. Awards will not be made solely on the basis of having saved a life.

The *Bronze Star Medal* (BSM) is awarded to any person who, while serving in any capacity in or with the Army of the United States after 6 December 1941, *distinguished himself or herself by heroic or meritorious achievement or service, not involving participation in aerial flight, in connection with military operations against an armed enemy*; or while engaged in military operations involving conflict with an opposing armed force in which the United States is not a belligerent party. *Awards may be made for acts of heroism, performed under circumstances described above, which are of lesser degree than required for the award of the Silver Star [Note: such awards are designated with a 'V' device for Valor].* The Bronze Star Medal may also be awarded for meritorious achievement or meritorious service.

The *Air Medal* (AM) is awarded to any person who, while serving in any capacity in or with the U.S. Army, will have *distinguished himself or herself by meritorious achievement while participating in aerial flight. Awards may be made to recognize single acts of merit or heroism, or for meritorious service.* Awards may be made for acts of heroism in connection with military operations against an armed enemy or while engaged in military operations involving conflict with an opposing armed force in which the United States is not a belligerent party, which are of a lesser degree than required for award of the Distinguished Flying Cross. *Awards may be made for single acts of meritorious achievement, involving superior airmanship, which are of a lesser degree than required for award of the Distinguished Flying Cross, but nevertheless were accomplished with distinction beyond that normally expected.* Awards for meritorious service may be made for sustained distinction in the performance of duties involving regular and frequent participation in aerial flight for a period of at least six months. In this regard, accumulation of a specified number of hours and missions will not serve as the basis for award of the Air Medal. Numerals, starting with two

will be used to denote second and subsequent awards of the Air
Medal.

The *Army Commendation Medal* (ARCOM) is awarded to any
member of the Armed Forces of the United States who, while serving
in any capacity with the Army after 6 December 1941, *distinguishes himself or herself by heroism, meritorious achievement or meritorious
service.* Award may be made to a member of the Armed Forces of a
friendly foreign nation who, after 1 June 1962, distinguishes himself
or herself by an act of heroism, extraordinary achievement, or meritorious service which has been of mutual benefit to a friendly nation
and the United States. *Awards of the ARCOM may be made for acts of
valor performed under circumstances described above which are of lesser
degree than required for award of the Bronze Star Medal. These acts may
involve aerial flight [Note: such awards are designated with a 'V' device for
Valor]. An award of the ARCOM may be made for acts of noncombatant-
related heroism which do not meet the requirements for an award of the
Soldier's Medal.*

Appendix B

RANGER[1] MEDAL OF HONOR RECIPIENTS

I would rather have that medal than be President of the United States.

—Harry S. Truman
Thirty-third President of the United States, 1884–1972

Name	Rank	Date	Unit
Nett, Robert P.	First Lieutenant	14 Dec 1944	Commander, Company E, 305th Infantry Regiment
Millett, Lewis L. Sr.[2]	Captain	7 Feb 1951	Commander, Company E, 2nd Battalion,27th Infantry Regiment, 25th Infantry Division
*Porter, Donn F.	Sergeant	7 Sep 1952	Company G, 14th Infantry Regiment, 25th Infantry Division
Mize, Ola L.	Sergeant	10-11 Jun 1953	Company K, 15th Infantry Regiment,3rd Infantry Division
Marm, Walter J.	Second Lieutenant	14 Nov 1965	Company A, 1st Battalion (Airborne), 7th Cavalry, 1st Cavalry Division (Airmobile)
Dolby, David C.	Staff Sergeant	21 May 1966	Company B, 1st Battalion (Airborne), 8th Cavalry, 1st Cavalry Division (Airmobile)
Ray, Ronald E.	First Lieutenant	19 Jun 1966	Company A, 2nd Battalion, 35th Infantry Regiment, 25th Infantry Division
Foley, Robert F.	Captain	5 Nov 1966	Commander, Company A, 2nd Battalion,27th Infantry Regiment, 25th Infantry Division
*Sisler, George K.	First Lieutenant	7 Feb 1967	HHC, 5th Special Forces Group (Airborne),1st Special Forces

Name	Rank	Date	Unit
Zabitosky, Fred M.	Staff Sergeant	19 Feb 1968	5th Special Forces Group (Airborne),1st Special Forces
Bucha, Paul W.	Captain	16-19 Mar 1968	Commander, Company D, 3rd Battalion,187th Infantry Regiment, 101st Airborne Division (Air-Assault)
*Rabel, Laszlo	Staff Sergeant	13 Nov 1968	74th Infantry Detachment (Airborne LRRP)173rd Airborne Brigade
Howard, Robert L.	Sergeant First Class	30 Dec 1968	5th Special Forces Group (ABN),1st Special Forces
*Law, Robert D.[3]	Specialist 4	22 Feb 1969	Company I (Ranger),75th Infantry1st Infantry Division
Kerrey, J. Robert[4]	Lieutenant (j.g.)	14 Mar 1969	SEAL Team 1
*Doane, Stephen H.	First Lieutenant	25 Mar 1969	Company B, 1st Battalion, 5th Infantry, 25th Infantry Division
*Pruden, Robert J.[5]	Staff Sergeant	29 Nov 1969	Company G (Ranger), 75th Infantry 23rd Infantry Division (Americal)
Lemon, Peter C.	Specialist 4	1 Apr 1970	Company E, 2nd Battalion,8th Cavalry, 1st Cavalry Division (Airmobile)
Littrell, Gary L.	Sergeant First Class	4 Apr 1970	Advisory Team 21 (Airborne Ranger)US Military Assistance Command, Vietnam
Lucas, Andre C.	Lieutenant Colonel	1-23 Jun 1970	Commander, 2nd Battalion, 506th Infantry,101st Airborne Division (Air-Assault)
*Gordon, Gary I.	Master Sergeant	3 Oct 1993	US Army Special Operations Command, Special Detachment-Delta, Task Force Ranger
*Shughart, Randall D.	Sergeant First Class	3 Oct 1993	US Army Special Operations Command, Special Detachment-Delta, Task Force Ranger

*Posthumously

1. Soldiers either assigned to a Ranger unit or a graduate of the U.S. Army Ranger School.

2. Earned his Ranger Tab in 1958.

3. First U.S. Army Ranger to be awarded the Medal of Honor while assigned to a U.S. Army Ranger unit.

4. U.S. Navy SEAL graduate of the U.S. Army Ranger School.

5. Only graduate of the U.S. Army Ranger School to be awarded the Medal of Honor while serving in a Ranger unit.

Appendix C

FORMATION OF THE 75TH RANGER REGIMENT

Never forget: the Regiment is the foundation of everything.
—**Marshall Wavell**
Field Marshall, British Army, 1883–1950

UPON THE COMPLETION OF the Arab-Israeli Yom Kippur War in 1973, the Pentagon grew concerned about the United States' strategic ability to quickly move well-trained infantry forces to any spot in the world. Even though the 82nd Airborne Division was considered "light," it still required a tremendous amount of airlift capability to get it to the fight. In the fall of 1973, General Creighton Abrams, Chief of Staff of the United States Army, issued a charter for the formation of the 1st Battalion (Ranger), 75th Infantry Regiment.

The battalion is to be an elite, light, and the most proficient infantry in the world. A battalion that can do things with its hands and weapons better than anyone. The battalion will contain no "hoodlums or brigands" and if the battalion is formed from such persons, it will be disbanded. Wherever the battalion goes, it must be apparent that it is the best.

The 1st Battalion (Ranger) was ordered activated on 25 January 1974, with an effective date of 31 January. Initially, Fort Stewart, Georgia, was considered home. In 1978–1979, the battalion moved to Hunter Army Airfield (HAAF) in Savannah, Georgia. To initially man this battalion and to serve as its nucleus, the men and equipment of Company A, 75th Infantry of the 1st Cavalry Division were transferred and the company eventually inactivated on 19 December

1974. The new battalion was assigned the heritage of the Vietnam era Company C (Airborne Ranger) 75th Infantry Regiment.

The 2nd Battalion (Ranger) was activated on 1 October 1974 and stationed at Fort Lewis, Washington. The men and equipment of Company B, 75th Infantry of the 5th Infantry Division Mech were transferred and the company deactivated on 1 November 1974. This second battalion was assigned the heritage of Company H (Airmobile Ranger), 75th Infantry Regiment.

In 1975, the Ranger black beret became only the third officially sanctioned U.S. Army beret. Literally centuries of lineage and battle honors were symbolized by the unique Ranger beret. Though little more than a dark piece of cloth to some, to those who serve as Rangers, the beret is representative of personal courage and selfless sacrifice. It is a symbol to be earned, not issued.

Unfortunately, the uniqueness of this distinctive symbol was impinged upon in 2001 when the U.S. Army Chief of Staff General Eric K. Shinseki made what was believed by many to be a misguided decision of having the black beret issued to all soldiers at large in an attempt to raise moral within the United States Army. When asked why the Ranger's black beret, Pentagon spokeswoman Martha Rudd was reported to have stated that the decision was based in part on fashion and that "it is the only color that goes with the Army uniform." A sad and insulting justification for the demise of such a cherished symbol. With the loss of its unique Black Beret, the 75th Ranger Regiment adopted the Tan Beret.

Following the invasion of Grenada during Operation Urgent Fury, the most recent Ranger battalion addition, the 3rd Ranger Battalion, 75th Ranger Regiment was reactivated on 3 October 1984 and stationed at Fort Benning, Georgia, with its World War II heritage to serve as its lineage.

To control the three Ranger battalions, a regimental headquarters was needed, thus leading to the formation and activation of the 75th

Ranger Regiment on the same day as the activation of the 3rd Battalion. The overall Regimental strength is approximately 2,000 Rangers. Each 580-man battalion consists of three 152-man line companies and a battalion Headquarters and Headquarters Company (HHC). Each line company has three line platoons and one weapon's platoon. Weapon's strength for each battalion is sixteen 84-mm Ranger Antitank Weapon's Systems (RAWS), six 60-mm mortars, twenty-seven M-240G 7.62-mm machine guns, and fifty-four 5.56-mm Squad Automatic Weapons (SAW).

U.S. Army Chief of Staff General John Wickham provided the first colonel of the Ranger Regiment—though technically and officially titled the third after the ceremonial commanders Brigadier General William Orlando Darby and Major General Frank D. Merrill of "Merrill's Marauders"—Colonel Wayne A. Downing, the following guidance on 10 May 1984 (author emphasis added):

> The Ranger Regiment will draw its members from the entire Army . . . after service in the Regiment . . . return these men to the line units of the Army with the Ranger philosophy and standards. Rangers will lead the way in developing tactics, training techniques, and doctrine for the Army's light infantry formations. The Ranger Regiment will be deeply involved in the development of Ranger doctrine. The Regiment will experiment with new equipment to include off-the-shelf items and share results with the light infantry community.

In the 1990s, another U.S. Army chief of staff, General Gordon R. Sullivan, would develop his own charter for the Ranger regiment with the following:

> The 75th Ranger Regiment sets the standard for light infantry throughout the world. The hallmark of the regiment is, and shall remain, the discipline and esprit of its soldiers. It should be readily apparent to any observer, friend or foe, that this is an awesome force composed of skilled, dedicated soldiers who can do things with their hands and weapons better than anyone

else. The Rangers serve as the connectivity between the Army's conventional and special operational forces.

The Regiment provides the National Command Authority with a potent and responsive strike force continuously ready for worldwide deployment. The Regiment must remain capable of fighting anytime, anywhere, against any enemy, and WINNING.

As the standard-bearer for the Army, the Regiment will recruit from every sector of the active force. When a Ranger is reassigned at the completion of his tour, he will imbue his new unit with tthe regiment's dauntless spirit and high standards.

The Army expects the regiment to lead the way within the infantry community in modernizing Ranger doctrine, tactics, techniques, and equipment to meet the challenges of the future.

The Army is unswervingly committed to the support of the regiment and its unique mission.

As previously noted, Special Operations had been granted the lineage and honors of the 1st through 6th Ranger Infantry Battalions of World War II in 1960. However, with the formation and activation of the Special Forces Operations Command, the lineage and honors of the Second World War Ranger Battalions, the Korean Conflict Rangers, and the Vietnam Conflict Rangers were rightfully transferred on 3 February 1986 to the 75th Ranger Regiment.

ACRONYMS

8ARC	Eighth Army Ranger Company	**GBU**	Guided Bomb Unit
AAR	After Action Review	**GWOT**	Global War On Terrorism
ABC	American Broadcasting Corporation	**HAAF**	Hunter Army Air Field
AC	Attack Cargo	**HHC**	Headquarters and Headquarters Company
AG	Assistant Gunner	**HLZ**	Helicopter Landing Zone
AGL	Above Ground Level	**HMS**	His Majesty's Ship
AH	Attack Helicopter	**HUMINT**	Human Intelligence
AO	Area of Operation	**IFV**	Infantry Fighting Vehicle
APC	Armored Personnel Carrier	**IOBC**	Infantry Officer Basic Course
ARCOM	Army Commendation Medal	**ISA**	Intelligence Support Activity
ARVN	Army Republic of Vietnam	**IV**	Intra Venous
ASP	Ammunition Supply Point	**JMPI**	Jump Master Procedures Inspection
BAR	Browning Automatic Rifle	**JOC**	Joint Operations Center
BCT	Brigade Combat Team	**JSOC**	Joint Special Operations Command
BDQ	Biêt-Dông-Quân	**KIA**	Killed in Action
BDU	Battle Dress Uniform	**LAW**	Light Antitank Weapon
C2	Command and Control	**LCA**	Landing Craft Assault
CAS	Close Air Support	**LCT**	Landing Craft Tank
CCB	Combat Command B	**LD**	Line of Departure
CCT	Combat Control Team	**LZ**	Landing Zone
CNN	Cable News Network	**MAD**	Magnetic Abnormality Detector
CP	Command Post	**MEDEVAC**	Medical Evacuation
CSAR	Combat Search-and-Rescue	**MH**	Military Helicopter
DZ	Drop Zone	**MIA**	Missing in Action
FARP	Forward Arming and Refuel Point	**MRE**	Meal, Ready to Eat
FLIR	Forward Looking Infrared Radar	**MSR**	Major Supply Route
FM	Field Manual	**NCA**	National Command Authority
FM	Frequency Modulation	**NCAA**	National Collegian Athletic Association
G2	Intelligence Section	**NCO**	Non-Commissioned Officer
G3	Operations Section	**NOD**	Night Observation Device

NVA	North Vietnamese Army	**SA**	Situational Awareness
NVD	Night Vision Device	**SATCOM**	Satellite Communication
OEF	Operation Enduring Freedom	**SAW**	Squad Automatic Weapon
OIF	Operation Iraqi Freedom	**SEAL**	Sea, Air and Land
OP	Observation Post	**SFC**	Sergeant First Class
OPCON	Operational Control	**SH**	Student Handout
OPLAN	Operations Plan	**SITREP**	Situation Report
OPSEC	Operation Security	**SNA**	Somali National Alliance
PA	Public Announcement	**SOAR**	Special Operations Aviation Regiment
PDF	Panamanian Defense Force	**SOCOM**	Special Operations Command
PFC	Private First Class	**SOF**	Special Operating Forces
PJ	Para-rescue Jumper	**SU**	Situational Understanding
PL	Platoon Leader	**TF**	Task Force
PLF	Parachute Landing Fall	**TO&E**	Table of Organization and Equipment
PO	Petty Officer	**TOW**	Tube-launched, Optically-tracked, Wire-guided
POW	Prisoner of War	**UDT**	Underwater Demolitions Team
PRA	People's Revolutionary Army	**UH**	Utility Helicopter
PZ	Pick up Zone	**UN**	United Nations
QRF	Quick Reaction Force	**UNITAF**	Unified Task Force
RAOC	Rear Area Operations Center	**UNOSOM**	United Nations Operations Somalia
RLTW	Rangers Lead the Way	**US**	United States
ROE	Rules of Engagement	**USS**	United States Ship
ROK	Republic of Korea	**VC**	Viet Cong
RPG	Rocket Propelled Grenade	**WIA**	Wounded in Action
RTO	Radio Telephone Operator	**WMD**	Weapons of Mass Destruction
S/SGT	Staff Sergeant	**WWI**	World War I
S2	Intelligence Section	**WWII**	World War II
S3	Operations Section		

INDEX

ABOUT THE AUTHOR

JOHN LOCK IS A 1982 graduate and former assistant professor of the United States Military Academy at West Point. He retired from active duty as a lieutenant colonel in May 2002. He enlisted in the Army as a private in 1974 and served as a non-commissioned officer until 1978. His commissioned assignments included the 1st Armored Division, West Germany, the 82d Airborne Division, Fort Bragg, N.C., Deputy Commander New York District U.S. Army Corps of Engineers and Chief Engineer Stabilization Forces (SFOR), Sarajevo, Bosnia i Herzegovina.

His military and civilian education includes the Engineer Officer Basic Course, the Infantry Officer Advanced Course, the Combined Arms Services Staff School, the Command and General Staff College and a Master of Science from Rensselaer Polytechnic Institute (RPI). His decorations include the Ranger Tab, Master Parachutists Wings and the Legion of Merit.

He is the author of *To Fight with Intrepidity: The Complete History of the U.S. Army Rangers, 1622 to Present* (Simon & Schuster/Pocket Books), *The Coveted Black and Gold: A Daily Journey Through the U.S. Army Ranger School Experience*, *Rangers In Combat: A Legacy of Valor*, and *Chain of Destiny*.

Lock currently works in support of architectural development, modeling, and simulation for the U.S. Army's Current Force and the Army's transformation to the Future Force, in addition to serving as a consultant in support of the Army Science Board and the Army's National Guard and Reserve senior mentor program.

His website is http://johndlock.com/
He can be contacted at JDLock82@aol.com

Printed in the United States
130836LV00001B/184/A